THE
RELIGION OF EVOLUTION

BY

M. J. SAVAGE,

AUTHOR OF "CHRISTIANITY THE SCIENCE OF MANHOOD."

**Fredonia Books
Amsterdam, The Netherlands**

The Religion of Evolution

by
Minot Judson Savage

ISBN: 1-58963-921-9

Copyright © 2002 by Fredonia Books

Reprinted from the 1876 edition

Fredonia Books
Amsterdam, The Netherlands
http://www.fredoniabooks.com

All rights reserved, including the right to reproduce this book, or portions thereof, in any form.

In order to make original editions of historical works available to scholars at an economical price, this facsimile of the original edition of 1876 is reproduced from the best available copy and has been digitally enhanced to improve legibility, but the text remains unaltered to retain historical authenticity.

TO

THE CHURCH OF THE UNITY;

WILLING TO BEAR THE PAIN OF THOUGHT, BRAVE ENOUGH
TO HEAR WHAT IS NEW, AND HAVING FAITH
THAT GOD WILL LEAD THE FREE
AND THE EARNEST TO
HIMSELF,

THIS BOOK IS LOVINGLY DEDICATED.

PREFACE.

In some form the theory of evolution is now accepted by nearly all the leading scientific and philosophic students of the world. It is rapidly giving its own shape to the thought of civilization. Science, art, human life, religion, and reform are becoming its disciples; and their tendencies in the near future must be largely determined by it.

Workers in many departments of thought have already reshaped their teachings into accordance with its principles; but so far as I know, in this country, no book has been devoted to a discussion of its effect upon religion.

This volume makes no claim to completion. It is only an essay in answer to the question, "If evolution is true, what have we left in the way of religion?" Some scientists affirm, and some frightened religionists exclaim, that evolution is essentially atheistic and irreligious; and that, if it is true, we have left no religion at all. The writer believes that it is the business of both science and religion to seek first and always for the *truth ;* for the truth only leads to God. He further believes that it is waste of time to seek to reconcile assumed truths. Truths are already at one, and need no reconciliation. Find and apply truth, then : the result is God's.

CONTENTS.

		PAGE.
I.	SCIENCE AND RELIGION	11
II.	THEORY OF THE WORLD	31
III.	THE GOD OF EVOLUTION	49
IV.	THE MAN OF EVOLUTION	73
V.	THE DEVIL; OR, THE NATURE OF EVIL	93
VI.	THE EVOLUTION OF CONSCIENCE	112
VII.	LOVE IN LAW	131
VIII.	PRAYER	150
IX.	BIBLES AND THE BIBLE	170
X.	THE DOCTRINE OF ATONEMENT	194
XI.	CHRISTIANITY AND EVOLUTION	215
XII.	IMMORTALITY	234

"Prove all things: hold fast that which is good."

PAUL.

"Let knowledge grow from more to more,
But more of reverence in us dwell;
That mind and soul, according well,
May make one music as before,
But *vaster*."

TENNYSON.

The truth-seeker is the only God-seeker.

The curse of both religion and science, in all ages, has been the thought that there was somewhere an ultimate,—a place to stop. Here we are, finite minds in the midst of infinity. And, for the finite that is moving toward infinity, there is nowhere a place to anchor, but only the privilege and the opportunity of endless exploration.

Beneath all the various, wide-spread, and disconnected labors, discoveries, and experiments of the great body of scientific workers, there is the common belief that all scientific truth is one; that the universe is all of one piece; that distant truths are only different parts of one divine pattern that runs all through the whole visible garment of God. This scientific faith is grander than any that the religious world has yet attained. But we must come to this. Religious truth is one, as God is one. Go forth, then, ye religious explorers, and seek only for truth; knowing that all truth-seekers are brothers, and must come to hand-clasping and looks of recognition by and by!

S.

"I apprehend that there is but one way of putting an end to our present dissensions; and that is, not the triumph of any existing system over all others, but the acquisition of something better than the best we now have."

CHANNING.

"IT is popularly said abroad, that you have no antiquities in America. If you talk about the trumpery of three or four thousand years of history, it is true. But in the large sense, as referring to times before man made his momentary appearance, America is the place to study the antiquities of the globe. The reality of the enormous amount of material here has far surpassed my anticipation. I have studied the collection gathered by Prof. Marsh of New Haven. There is none like it in Europe, not only in extent of time covered, but by reason of its bearing on the problem of evolution. Whereas, before this collection was made, evolution was a matter of speculative reasoning, it is now a matter of fact and history as much as the monuments of Egypt. In that collection are the facts of the succession of forms, and the history of their evolution. All that now remains to be asked is *how*, and that is a subordinate question." — PROF. HUXLEY, *before the American Association for the Advancement of Science, August, 1876.*

THE RELIGION OF EVOLUTION.

I.

SCIENCE AND RELIGION.

THE phrase, "the conflict between religion and science," has become a very common one in newspaper, in magazine, in public address, and in sermon; and it represents the observed fact that there is a grand division running through the thinking minds of civilization, on one side of which stand the advocates of religion, and on the other the advocates of science. Not, by any means, that there are no religious men on the scientific side, and no scientific men on the religious side, but that this division does represent a real and general fact, and that these two sides stand in a certain antagonism to each other. But yet, strange as it may seem, I suppose it to be still true that there is not a scientific man living who would claim that a real truth of science can by any possibility come into conflict with a real truth of religion; and there is not a religious man living but would confess that it was simply an impossibility that a real religious truth should stand in antagonism to a real scientific truth. But the antagonism of the attitude still remains, because it is true that on both sides there are large bodies of men that are very much more concerned in establishing their positions than they are in finding out what is the truth. It seems to me a very strange thing that any man should be willing to hold such an attitude as this, either on the side of science or religion. There is no possibility of its being for the permanent interest of any man that he should be able to establish himself in a falsehood; for though he may build him a house as wide as the earth, and as high as the heavens, if its foundation be in the sand, the floods of the eternal movements of the divine forces will some time undermine and sweep it away. There can, then, be no interest in any man's holding to a position that is not true; so that one might suppose that the chief anxiety of men would be, not to prove that they were right, but to find out whether they were right. And yet I have met men among my own personal friends, who would tell me candidly, and with their whole hearts, that so dear to them had become the positions that they had inherited, even if they were false, they did not wish to find it out; they did not wish to be disturbed; they did not wish to be compelled to re-adjust their thinking to any new truths: for it is one of the inevitable facts of the world that a new truth is an "unsettler" everywhere. It comes in to disturb and to shake old institutions, and to demand of men that they do not build forever their house in the place where their fathers builded, but that they regard it simply as a tent, to be folded and taken with them on a forward march towards something which is higher and grander and broader in the way of truth.

This fact to which I have alluded — that there is this antagonism between men on account of their being anxious to establish their own positions rather than to find out truth — I suppose to be true in a larger degree of the defenders of religion than it is of those that stand on the side of science; and I conceive that there is a very natural and, in one sense, a satisfactory reason for it. You cannot make any scientific man feel anxious about any supposed scientific truth on the ground that its truth or falsity will endanger the welfare of his soul, either in this world or in any other world. But when a man has inherited some religious belief that is intertwined with all the sacred associations of the past, with the present affections of the soul, and with all the dearest and grandest hopes of the future, it seems to him, when you touch it, that you are unsettling the universe, that you are sweeping away from him every thing that is dear, every thing on which he has been accustomed to rely. You take away the anchor of his soul; you cut the cable which bound him to any sure hope and abiding-place, and he is set adrift to float, nobody knows where.

I say, then, that it is natural to a person who has had a training like this that he should be jealous of the incoming of something that claims to be scientific truth, that conflicts with what he has been taught to regard as religious truth.

Now, in order that we may understand something of the principles of this conflict that has been going on since the dawn of civilization, and in the midst of which we are still engaged, — in order that we may understand something of the methods of it, and that we may be able to forecast with some degree of probability the outcome of the battle of the present hour, it will be needful for us to go back, and glance for a moment at a few of the fields that have been fought over and won in the past.

Religion held a universal sway over the mind of man before science was even born; for religion is as old as the instincts of hope and fear in the human soul, and has bound itself up with these hopes and fears; so that this conflict has not simply arisen under Christianity: it is older than Christianity. I say this because some scientific men speak as though Christianity, and no other religion, was the grand obstacle that had stood in the way of scientific progress. It is not because Christianity is any different in this respect from any other religion in the world, but simply because Christianity happens to be the religion of a civilization where this conflict has been going on. But the conflict began before Christianity was born. The old Greeks supposed that the sun, in his grand march across the heavens from the east to the west, was a god driving his flaming chariot; and they worshipped this god with incense and with temples and with offerings; so that his ritual was widespread all over the ancient world. When, then, some thoughtful philosopher came forward first, and, as the result of his study, dared to broach the heresy that the sun was no god, after all, but only a ball of flaming fire, he was unsettling the foundations of the religion which was dear to the popular heart; and the people resented it, and fought against it with just as bitter a feeling of opposition as that which actuates the hearts of the theological defenders of what they claim to be essential religion at the present day.

But the first great battle between the advocates of science and the advocates of religion was that concerning the geography of the world, — the question as to whether it was round, or whether it was a flat surface. It seems strange and incomprehensible to us, to-day that it could possibly make any difference to the advocates of religion whether the world was round or flat; and yet one of the bitterest contests of the world raged over this question for ages. And so high did the feeling run, and so bitter was the opposition on the part of the priests of the Catholic Church (and the ministers of the Protestant Church, as well, — for they were linked together hand in hand in fighting that battle), that one of the priests of the middle ages went so far as to say that the Church could better endure having the existence of God called in question, or the immortality of the soul, or the religious nature of man, than that it should listen to the damnable heresy that the world was a globe, and not a flat surface. And Luther and Melanchthon, those grand lights of the Reformation, went quite as far in their opposition to this new science as the priests of the Catholic Church. And what were their arguments? Why, such as these: That the Bible spoke everywhere of "the face of the earth," and said nothing about any other side but the face. Again: that, if there were any antipodes living on the other side of the world, then the character and government of God were impeached, because he had made no provision for their salvation. And again: the command had been given to the apostles to go into all the world, and to preach the gospel to every creature; and since the apostles had never visited any nations at the antipodes, therefore there were no such nations. These were their arguments, — arguments brought from a superficial understanding of the biblical use of language. And they went so far as to construct their theory of the world after the precise pattern of the Jewish tabernacle in the wilderness, saying that that was the divine copy of the universe; and that, as the tabernacle was twice as long as it was wide, and had just such and such proportions, therefore the universe was in just this same shape, — an oblong square; and that the world was supported by pillars, as the tabernacle was supported by pillars; and they invented some sort of a grand mountain at one end of this oblong square, behind which the sun was pulled at night when it was dark. So far did they carry this battle over the question as to whether the world was round or flat. And one thing which seems very remarkable to us who are accustomed to go to Nature and interrogate her, when we wish to get an answer, is, that it never seemed to occur to these men to find out whether the world was round by travelling over it, by sailing around it, by measuring an arc of its surface to see whether it was curved or not. It never seemed to occur to them to go to the natural phenomenon itself, and ask it the question. Instead of that, they went to certain old books, that they regarded as sacred, to find out what somebody who lived thousands of years before had thought about it. This, then, is an indication of the battle that was fought over the question of geography.

The next great conflict was over the position of the earth in the solar system, — as to whether the earth was the centre, or the sun was the centre. And here, again, instead of looking to find out whether their theory was true, instead of prosecuting the study of astronomy, instead of opening their eyes, they seemed to think that to dare to scrutinize the works of God was impiety and heresy; and so science was fought against with every weapon, not only of the civil power in the way of persecution, but with the most opprobrious epithets and with social ostracism. And here, again, the arguments are very strange that they bring. Luther laughed at and ridiculed the foolish scientific men of his day who said that the sun was the centre of the solar system, and that the earth revolved about it, and clinched his argument by saying that Joshua commanded the sun to stand still, and not the earth; and therefore it could not possibly be that the earth moved around the sun, instead of the sun moving around the earth. And then, when Galileo invented his telescope, so that he could see the moons of Jupiter, instead of looking through his telescope, and finding out whether he really did see the moons, they charged him with being in league with Satan, and said, that, through Satan's help, he had invented an instrument which created the heavenly bodies which he claimed to see. And here, again, with these old ideas inherited from the past, they fought against finding out what were the real facts in the realm of nature concerning the way in which God had constituted this wondrous universe of ours.

The next battle that I shall speak of (I dwell on these very lightly, simply by way of illustration) is one that you are familiar with yourselves, that has been fought out in the lifetime of almost every one that has attained to the age of twenty-five years, — the battle over the antiquity of the earth and of man. The battle started in the new discoveries of geology. Here, again, the same old weapons were used in the fight against this new, grand, and higher truth. When sea-shells and the fossil bones of fishes were discovered upon the sides, and near the summits, of high mountains, instead of believing what geology taught, — that the mountains had themselves been under the sea, and had afterwards been raised by the natural action of the forces of the earth, — they claimed that the presence of these things must be explained upon the theory that the flood had carried them there, and had left them behind when it had passed away: or they claimed that these were false creations of Satan, made as an imitation of, and a parody upon, the works of God: or they claimed that they were the first attempts of God in the way of creation; that he had to try several times before he succeeded, and that these were the remnants of his failures; that he had thought out better ways afterwards, and created the existing specimens of life on the face of the earth after newer and finer patterns. And at that time the grand obstacle to the reception of the new and magnificent truths that geology taught was simply the old interpretation of Genesis. Because Archbishop Usher had ciphered it out that the world was only six thousand years old, therefore geology could not be true; and because of the institution of the sabbath, based on the supposition that God worked six days, was then tired, so that he wanted to rest on the seventh, and therefore set apart that day, — this was again used as a conclusive argument against the facts that could be seen by simply looking at them. But this battle is at last fought out and ended; and the antiquity of the world, and the antiquity of man, reaching back, not six thousand years, nor ten thousand, nor twenty thousand, but possibly, in the case of humanity itself, a hundred thousand, certainly in the case of the world thousands of thousands, if not thousands of millions, — this question is at last absolutely settled for the mind of every man who is at all familiar with the facts that go to prove it.

And then there is one battle more that I will glance at, — one which is raging very furiously at the present time, represented on one side by men like Herbert Spencer, Darwin, Huxley, Tyndall, the German Haeckel, and on the other by the leading men of the theological world; and this battle is nothing more nor less than concerning the methods of creation, and what was the origin of life upon the globe, and by what process living creatures have developed from the first simple beginnings in the primeval oceans up to the grandest manifestations of the intellect of humanity. This is the battle in the midst of which we are living to-day. I shall say nothing in regard to the merits of it at this time, because it does not come in as a necessary part of my subject.

But now let us look for a moment, and see what are some of the principal results of these battles between religion and science. And, in the first place, I should think it would be a very discouraging fact for those who are so afraid of science, that, in every single one of the grand conflicts of the past, the advocates of the popular religion of the time have been beaten out and out; so that to-day there is not a remnant left of them. There has not been a well-thought-out and well-arranged contest between the advocates of science and the advocates of religion in the history of the past, in which the advocates of science have not been completely and permanently triumphant. What is the next result? The next result is, that religion, so far from receiving any detriment on account of the overthrow of those that have assumed to be her champions, has grown grander and more magnificent every time. Religion, in other words, has been helped, advanced, uplifted, magnified, and made grander by the conquests of science. And how? Not that certain definite forms of religion, certain theories of theology, certain sectarian claims, have not been injured; for these have been overthrown, and ground to powder. But these are not religion: these are simply the false and mistaken theories of men. These have been trodden down in the track of the advancing thought of humanity; but religion has been made grander. Just think of it for a moment! The God of that little universe which was patterned after the Jewish tabernacle was simply a gigantic and non-natural man. He was simply the autocrat of a little kingdom not so large in the conception of the world, at that time, as the United States is to-day; simply a sovereign ruler, who could make the mountains tremble with his footstep, the rumbling of the thunder being the roll of his chariot-wheels, or the threatening tones of his voice, while the lightning was the gleam of his sword, or the flash of his eye. Think of a God like that!—not grander than the Olympian Jupiter. And now words and time both fail me to do more than ask you to think of the infinite enlargement of the conception of God that has come from the revelations of science, — the world a magnificent globe, though the smallest of all the worlds with which we are familiar, yet sweeping around the sun in its wondrous orbit; and this system, mighty and grand as it is, one of the very least with which we are acquainted, and the universe made up of a countless number of unspeakably grand systems and galaxies that stretch off and off and off, until we know not whether this material universe of ours be not itself absolutely infinite and unbounded. Think of the difference in the human conception of the God that resides at the centre, who is the life and force and power and beauty and glory of a universe like this. And this is the object of our religion, that the advance of science has given us, — an advance right in the teeth of the defenders of religion, and made at the cost of overthrowing and trampling down that which the religious advocates of the time regarded as essential to the very existence of religion itself. This is an indication, an illustration, of how science has exalted our conception of God, of the universe, of the nature of man and the grandeur of human destiny; so that religion has been an unspeakable gainer by being defeated in its conflicts with science. And yet there has resulted one incidental evil, — an evil not growing out of the fact that religion was defeated; for it would have been the worst possible thing for humanity if religion had gained in this contest with the science of old: but this incidental evil has followed from the simple fact that religionists put themselves in the way of the advance of truth concerning God and his universe. And how has the evil been wrought? It has been wrought in this way: Scientific men have come to look upon the church and theology as simply superstitions, outworn obstructions in the pathway of the progress of discovery. And they have been justified in regarding those who have put themselves forth prominently as the advocates of religion as obstructionists in the way of the advance of truth; for every single time that there has been made a proposal to enlarge the kingdom of definite knowledge, there has been the same old tiresome outcry of "Science falsely so called!" and the opposition of religion, on the ground that they were driving God out of his universe. Why, when Newton discovered the law of gravitation, that which all ministers now for a hundred years have been speaking about as one of the grandest illustrations of the power and greatness of God, — when Newton first made that discovery, he was branded as an atheist and an infidel by the Church, because, as they said, he had taken the universe out of God's hands, and had given it into the keeping of a law. And so every time that there is a new law discovered, the same old cry is raised, until to-day the crowning objection that is urged against the law of evolution is, that it is driving God clear out of his universe; as though a law were any thing more than simply a name for a method of divine working. And the broader you make the law, the more comprehensive and sweeping, the longer its reach, only the more grand and magnificent become the thought, and the conception of a God who is able thus to weave a network of law that shall cover the universe, and reach through all time.

And now let us stop for a moment, after reviewing these contests, and let us dare to look this thing called science in the face. What is science? Why should religious men be afraid of it? That they are afraid, and that even liberal men and women are afraid of it, I have found out by personal conversation with some of my own parishioners and friends. And there is a sort of uneasy feeling on the part of many, that when a minister says "science" in the pulpit, he has somehow temporarily dropped his real vocation; that he has left the work of preaching God's truth, and is talking about man's theories and ideas; that he is really trenching on the ground of secularism, doing work appropriate for Monday, but not for Sunday. Let us, then, I say, for a moment look this matter of science in the face, and see if we can find out what it is. Science is nothing more nor less than the arranged, organized, definite, verifiable knowledge of the world. It is nothing more fearful than that. We know, for example, that a stream of water cannot possibly rise higher than its fountain. Having established that as a law of the movement of water, we have established so much as being a part of science. Again: we know, as another example that certain chemical elements brought together in solution will arrange themselves, under certain conditions, into a crystal of a certain form ; that they will do it every time. There, again, is simply another fact of science. We know now, since the battle is fought out and won, that the earth turns around on its own axis once in twenty-four hours, and that it revolves around the sun. These, again, are simple facts of science. That is, any thing concerning man; concerning human society; concerning the earth; concerning the animals that inhabit the earth, the grasses, the herbs, the flowers that spring out of its soil; concerning the clouds that sweep through the air; concerning the atmosphere itself; concerning the sun and the stars, — any thing that we know and can prove about any of these things is science. It is real science when it can be verified, and when it is not somebody's guess or supposition. Now, there are a great many things that are simply believed, a great many theories that are called hypotheses. They are assumed as the working implements of science. They are not science until they are established. Things that are proved, and that can be proved again, are science. And here is one grand advantage that scientific truth has over any other, — that which is true in science is true everywhere, and it is true always. I do not accept it on the basis of a man's dream who lived five thousand years ago; I do not accept it because somebody comes to me, and says that he talked with an angel, and the angel told him so; I do not accept it because some man says, "I was inspired to write such and such things, and this was one of the things that I was told to write;" I do not accept it on testimony from anybody. It is a truth of the living, working God, right before my eyes to-day; and I can prove it now, and I can prove it to-morrow, just as well as it was proved yesterday; so that there is no possible chance for contradiction or conflict as to whether it is true or not. That is the one grand thing that is claimed in favor of the methods and facts of science.

And now, is this science sacred, or is it secular? There is no possible way by which I can absolutely settle it that every part of the Bible was inspired by God. There is no possible way by which I can find out how much of the Pentateuch Moses wrote, or whether he was inspired to write that which he did write, or whether he wrote it out of his own human wisdom. There is no possible way, I say, by which I can establish questions like these. And yet these things by many, by the multitudes of Christendom to-day, are held out before us as the things that are peculiarly and especially sacred; and our reverence is demanded for them. But whether I know any thing else in the world, if there be a God at all, I know that he is the author, the life, the force, the beauty, the glory, of this universe that he has made. And when I look upon some little flower bursting through the sod, I am looking directly into the secrets of God's beauty and God's taste. And when I look into the eyes of my friend, and when I see the self-sacrifice, the self-denial, the love that is prompted by his heart, I am looking at a fact of science, something that I can recognize and prove as a part of human nature; and I am at the same time looking into the divine mystery of the love and the self-sacrifice of God. When I look at the stars coursing their ways through the blue deeps of heaven, I am, as Kepler said of old, "thinking over after him again God's own thoughts." And so, anywhere where God has been, or where God is now (for he is now where he ever has been), wherever God is, I look upon his very footstep, and I can put my finger into his own finger-prints; and I can see God's life in the growth and progress of nature about me; I can feel the divine pulsations in the air, and in the life of my body; I am living in the midst of the only temple that God himself has consecrated, and that I can be absolutely sure is a representation of God's own work. Whether there are mistakes about any thing else or not, this is certain. Here, then, in nature, — in sun and star, and sky and cloud, and ocean and earth, and grass and flowers and trees, and human nature, — I am looking directly into a revelation of God ; and if I can read it, I can read the very thoughts and processes and methods of the divine working and development.

This, then, is science, that men dare call irreligious; that men dare decry in the name of human traditions, and human systems, and human dreams, and human follies, and human superstitions. If there be any one thing that is the sacred book and teaching and life and outcome and revelation of God's own heart and God's own power, it is that which is called science.

And now, in just a word, what is religion ? and why need religion be afraid of science ? and what ought to be the real relation existing between science and religion ? Religion, — I am going to give you a very short definition ; and yet, if you will think about it as long as you will, I do not believe you can find any thing that reaches beyond its limits, — religion simply means the relationship, as to right or wrong, in which man stands to his God and to his fellow-man. This is the whole sum and substance of religion. It covers and includes it all.

Now, then, what can science do for this religion ? Science has been doing for hundreds of years one of the greatest services possible. It has been destroying the superstitions, the crudities, the falsehoods, the misconceptions of men concerning religion. For example, the doctrines of astrology, of demoniacal possession, of witchcraft, the doctrine of the material resurrection of the body, of a material hell just under the surface of the ground, and many others that were once considered central and essential parts of religion, — these things which were only hurts and damages, barnacles on the ship that hindered its sailing, — these things science has stripped off, and thrown away, and utterly destroyed.

I do not wonder that men have cried out against science because it has done these things; for if once a man identifies his own thought with the very central life and thought of the universe, of course, when you touch him, he thinks the throne of God is giving way. But science has reconstructed religious thought: that is one thing that it has done for it. Another thing I have already enlarged upon. It has heightened infinitely the objects of religion, giving us a grander God, a nobler humanity, a more magnificent universe as the theatre for human action. Another thing: if I am to understand definitely what is right and what is wrong, if, in other words, I am to be intelligently moral, if I am to educate my conscience after the true pattern, I must learn definitely what are the facts concerning human nature, concerning its origin, concerning its history, concerning its relationships. I must study out these, and definitely settle them; and in that way alone can I be sure that I am living a truly moral life. For example, as illustrating what I mean: the larger part of the religion of the past has been simply ritual. It has been said, "You must obey the priest, you must go to church, you must pay tithes, you must partake of the sacraments, you must do this thing, you must do that thing." Now we believe, and, in the light of science, it is perfectly clear to us, that none of these things have any essential relation whatever to religion; and a man may be religious, and disregard every one of them, and a man may do every one of them, and a thousand more, and be utterly and through and through irreligious.

Science, then, is definitely settling for us what is right and what is wrong, by observing the facts in regard to human nature.

And, then, one thing more, science is gradually giving to religion its methods,— its method of study and of proof. Why is it that intelligent and thoughtful men to-day are disputing so largely the great central questions, as they are claimed to be, of religion? It is one of the strange things of the world, if the claims of the Church are true, that thoughtful, intelligent, earnest, religious men all over the world are disregarding them. Why, no man thinks of disregarding the law of gravitation, because he knows he would be a fool to do it. Whether he disregards it or not, it is a grand fact of the universe, and, if he comes in its way, it will crush him. No man disregards the laws of fire, because he knows that fire is a fact of the universe that it will not do to disregard. No man questions as to whether the earth moves around the sun, because it can be proved that it moves around the sun. But the reason that religion is so disputed, that it is so bandied about, that it is so fought against, is, that theology has sought to identify with religion a hundred or a thousand things that no man on earth can prove. As an illustration of what I mean, Mahomet claimed to be inspired of God, and to teach the doctrines of the Koran under inspiration; Swedenborg claimed to be inspired of God; Peter and Paul claimed to receive visions direct from God. Now, whether they did or not, nobody can tell. There is no possible way of proving that they did or did not. In other words, to state very definitely and very clearly what I mean, many of the asserted truths of religion have been heretofore "all in the air,"— truths that nobody could touch, nobody could feel, nobody could see, nobody could prove; so that you could accept them if you wished to; but you could not make it a rational necessity for another man to accept them. But once bring the methods of science into the sphere of religion, establish the existence of God; establish the laws of God as manifested in the universe, establish the nature of man, and his religious nature as a part of it, — establish these on the firm foundations of investigated and verified truth, and no reasonable man will then think of questioning them, or questioning whether he ought to be religious, and lead a religious life. He will no more think of questioning these than he will the fact that the earth moves around the sun.

And so science more and more ought to come into the sphere of religion, and bring with it its methods of investigation and proof. And so, as the result of the present conflicts, there is coming a grander revelation of religion, and an establishing of it on immovable foundations, such as heretofore the world has never seen.

II.

THE THEORY OF THE WORLD.

At the National Unitarian Conference in Saratoga (1874), one of our most widely known ministers was making a speech on our missionary work; and, in the course of his remarks, he took occasion to speak slightingly of those who were wasting their time on such unpractical questions as the antiquity of the world and the origin of things. He thought there were enough problems of real, pressing, living importance right about us to absorb all our attention, and consume all our energy. And a year ago (1875), in Music Hall, in the course of a lecture on "Our Scandalous Politics," Mr. Parton took occasion to ridicule those who were troubling their brains over the theories of Darwin and Spencer, instead of grappling vigorously with the political evils and social reforms of the day.

Thoughts and utterances like these are natural enough to one who does not look beneath the superficial movements of the time for the hidden, and oftentimes remote, springs and causes of the conditions of things. If one knew nothing of the interior mechanism of a watch, he might think he could make it keep good time by turning the hands round on the face; but a wiser person would take it to a watch-maker, and have the origin of the outer movements looked after.

Men who pride themselves on their reputation as "practical" workers are often very impatient of theories and theorizers. Theorist to them means visionary. They regard him as dwelling in a cloud-land, and dealing with unsubstantial fancies. They think his fitting representative is the fabled dog on the foot-bridge, who dropped his bone while clutching at a finer looking shadow. They propose to hold by the bone. And yet, if you'll think of it, every man living has his theory of every thing he does; and all his practice is the result of his theory. The farmer who sneers at his neighbor for adopting the "new-fangled notions,"— the knowledge of chemicals and soils that modern science has revealed to him, — and who pins him to the walls of his kitchen with the stigma "theorizer!" is himself a theorizer just the same; only he keeps working away on a theory that consists in bungling tools, — and guessing experiments, and back-breaking hand-labor, — a theory that kept his grandfather poor, a theory long since exploded as deficient and half-way, — instead of accepting that theory that new investigation and successful practice have proved true. Should a boy at school attempt a problem in mathematics, and pay no attention to the theory, the underlying principles according to which the question could be solved, you would say he ought to exchange his seat for the dunce's block. Watt and Stephenson succeeded in laying the foundations of our present railway system, when they discovered the true theory of the laws and application of steam. Von Moltke is the greatest of modern generals, because he has the brain to conceive and carry out the most nearly perfect theory of the laws of war. All of us are theorizers who have brains enough to think out, and work along, the lines of any efficient plan in our business. The banker, the merchant, the lawyer, the physician, the minister, each has his theory; and he is successful just according to his ability to discover the true theory of his position, and to carry it out in effective practice. So you might as well talk of practically growing an oak without an acorn as to think of successful practical work divorced from theory.

And now, for a moment, glance at the absurdity of the position assumed by the preacher and lecturer just referred to. Society and state are sick with various maladies which they desire to heal: so, without stopping to waste time and strength on the unpractical questions of the remote origins and causes of the disease, they propose to blister and bleed and cauterize, without any loss of time. An intelligent physician, when called to a patient, does not consider it any waste of time to stop and investigate, and study the symptoms, in order to find out what the matter is; and, the more serious and urgent the disease, the more careful he is to do this.

Now, there is not a single personal, social, political, or

religious question of the day that does not run back and
root itself in the remotest antiquity of the race. They all
grow out of the original nature of humanity, just as the
topmost twig or leaf of a two-thousand-year-old tree is
the outgrowth of the first germinal principle from which
the ancient trunk has sprung. Here is the trouble with
most of the "reforms" of the age. They are the outcome of transcendental notions, purely empirical study, or
the hasty guesses of enthusiastic persons, who propose to
finish in a year a structure the foundations of which God
has been centuries in laying. Any true reform must
know the drift of the ages, and work in the line of the
eternal movements of the universe.

We are now prepared to raise the question as to
whether a study of the "theory of the world" is a practical matter. And, in the first place, glance at the facts.
All nations on the face of the earth who have been
civilized enough to have any thought-out and organized
religion have always connected their popular religion with
a cosmogony, or theory of the origin of the universe. The
character of their gods, their conception of humanity,
their codes of morals, the rights of rulers and subjects,
their hopes and fears of a future life, have all been the
outcome of their conception of the universe. Their whole
practical life has been the simple result of their theory of
the origin of things.[1]

And how is it with Christendom to-day? The popular

---

[1] Carlyle has said, "Tell me what a man thinks of this universe, and I will tell you what his religion is."

conception of the nature and attributes of God, the nature of man, the origin and nature of evil; the practical question of sin; the ecclesiastical schemes of salvation, heaven and hell; the prevailing theories of government and of social progress; the status of woman; the rights of children, — all the great practical questions of humanity are the direct outgrowth of the Mosaic cosmogony, the Jewish theory of the origin of the universe. And the present condition, together with the past battles and progress of science, is the natural result of this same cause. Not practical, or of present importance? There is not a single question of the age, that for present, practical, pressing importance, begins to approach the one that Spencer and Darwin and Haeckel have raised. You might as well say that because the sun is ninety-two millions of miles away, its influence, and the laws of its life and shining, are of no practical importance to Boston. Boston is an outgrowth of the sunshine, from the granite that paves its streets to the brains that rule in its counsels. So the present active world, with all its widespread and multiplied interests, is the outgrowth of the far-off origin of things.

To specialize a little more particularly, and let you see how intimately religion is connected with the theory of the world, I ask you to look at the Church of the last two thousand years. Please observe that the whole orthodox system is the natural and logical outgrowth of the Mosaic account of the beginning of things in Genesis. The prevailing beliefs about God, the nature and fall of man, total depravity, the need and the schemes for supernatural redemption, the whole structure, creed, and ritual of the Church, the common belief about the nature and efficacy of prayer-meetings, the whole system of popular revivals, limited salvation and everlasting punishment, — every single one of them is built on the foundation of the Mosaic cosmogony. And there is not one of them all but will be destroyed or modified when it shall become popularly settled that the Mosaic cosmogony is not a correct account of the facts.

Having made it appear, then, that as practical, earnest men of to-day, it is well worth our while to look into and investigate this question, I now ask you to go with me to a consideration of the only theories that need detain us. Of course we need not stop even to glance at the fantastic notions that prevailed among so many nations in the childhood of the world. Only two theories, the Mosaic and the Evolution, even pretend to claim the sober belief of our nineteenth-century civilization. To these, then, we must confine our attention.

I. The Mosaic Cosmogony.

Before going any farther, I wish to make two or three remarks that are worth careful attention.

(1) The account of creation in Genesis holds its place in the popular belief, not because it has been proved, or is capable of proof, but solely because of its supposed necessary connection with the truths of Christianity. This is, at any rate, a strange and questionable basis on which to found a scientific belief.

(2) It is an old-world traditional belief, not for the first

time revealed to Moses, but one that came down from a time long before the foundation of the Hebrew nation.

(3) It does not even claim to be the result of a study of the facts that it proposes to explain. No such study was then possible, or had even been attempted; so that Moses is not telling what either he or any one else at that time had any way of knowing. It is only the traditional belief of that age.

(4) These traditions get a great deal of hardly deserved reverence and belief from the fact of their high antiquity; just as a man is proud of his ancestry, though the roots of his family-tree run down into outright barbarism. But, if you'll think of it, the reverence belongs here. We are the real ancients. The present is the hoary antiquity of the earth. 'Tis a man's old age, and not his childhood, that wears wisdom and gray hairs. This story of Moses is one of the fancies of the world's childhood. Never was civilization so old, and never had it such stores of accumulated knowledge, as now. In fact, never, until within the last hundred years, has the world gathered enough about the facts of the universe, so that mankind was competent to frame a reasonable theory of the world out of its acquired knowledge. If, then, in the history of humanity, there has ever been a time when there was a possibility of settling this question, now is that time.

(5) The Mosaic cosmogony has no scientific claim to be called a theory at all; for the simple reason that it explains nothing whatever. Its very claim to be an explana-

tion is merely a leap into the incomprehensible. It simply says, "God made it." But that does not at all explain the method of creation,—how, by what process, and according to what laws and forces, things have come to be as they are. I do not explain the mystery of life when I tell my child that God made the new baby, and that the angels brought it down to me. I do not explain a tree, when I tell a child that God made it. I do not explain a question in arithmetic, when I tell a pupil that I worked it out, and that so and so is the answer. But all these are explanations just as much as Genesis explains the world. Men seek the causes and the methods by which results have been produced. The solar system is explained by the law of gravitation, not by saying God or the angels make the planets move.

Thus it is perfectly safe to say that no one would think of resting in the Mosaic story, were it not supposed to be a part of their religion to do so. And evolution has been opposed, not because it could not give good reasons for itself, but because it has been regarded as hostile to the popular religion.

With these remarks, then, in mind, we are ready to look at Genesis. The popular belief has been simply this: God had lived alone, complete and happy in himself, from all eternity. Suddenly, for no conceivable reason, he concluded to create the world. Previously, however, he had made the angels, that they might serve and praise him; though what service or praise he needed, who was complete in himself, it were hard to tell. If he had any motive in creating the world, it was thought to be that he might glorify himself, and receive the admiration of his creatures. The chief result was to be, that, after the world had passed away, his goodness in saving a few, and his justice in damning the many, might be seen as the result of his scheme of redemption. When, then, he was ready, by the word of command, he created matter out of nothing, and of this matter built the world. This world, a flat surface, he anchored in the midst of space, "setting it fast forever, so that it could not be moved." Then he elaborated the solid concave arch of the firmament, and placed it like a dome over the earth. In this he arranged the sun, moon, and stars, to divide the seasons, the days, and the nights, and to give light to this earth. He separated the waters, making the oceans of that which he left on earth; and in his storehouses "above the firmament," he treasured up the rains to water the earth, which watering was to be done by opening the windows of the sky, and letting the water through. Then he made the different forms of life, creating fishes and reptiles, and birds and animals, out of the dust. When this is done, it occurs to the Deity that none of these creatures can think of or praise him: so he consults, and concludes to make man in his own image. He forms Adam out of the dust, and then breathes in his nostrils, and he becomes alive. Then, seeing that he is lonesome, he concludes to make a woman to keep him company. So he puts him to sleep, takes out one of his ribs, — which, strange to say, has never been missed. — and from it constructs Eve. All this has taken him six days. He is now tired, and gives one day to rest. This is the origin of the sabbath. If by resting is meant letting creation alone, we should suppose it might have fallen into disorder during the neglect. But if it means simply ceasing to create, then he has been resting ever since, on this theory; and it is hard to see why it is stated that he took only one day. But we know now that the process of creation has never ceased: so that we can get no meaning out of it at all. Even Jesus declared that his Father was working still.

Now, it is not my intention to insult your intelligence by proving to you that this is not true. The conception of God and of his methods is such as the world's childhood might ignorantly imagine; but no free intellect of the present age cares even to refute it. Only to look at it is sufficient refutation. Genesis contains contradictory accounts even of the original creation; and the inspiration of the Old Testament was not such as to prevent the most palpable mistakes being made in describing natural things.

We may now pass on to consider,

II. The Theory of Evolution.

It would transcend our limits to attempt even an outline of the proofs of this theory. These are to be found in the works of the masters of science, specially prepared for that purpose. I must, therefore, content myself with remarking some of the surface probabilities, and then placing the theory itself alongside the Mosaic, that you may compare them.

(1) It is a fact that ought to make men stop and think, before rejecting it, that almost every trained scientific man living, who is competent to give a judgment on the question, is a believer in evolution. If all the skilful doctors were agreed about a certain disease, it would hardly be modest for us to say they were wrong. When all the generals are at one about a military question, the probabilities are decidedly their way. When all the architects agree about a building, and when all the painters unite in defence of a question in art, outsiders should at least hesitate. Nearly all the present opposition to evolution comes from theology; but theology does not happen to know any thing about it. As though I should attempt to settle a disputed point in music by the sense of smell, or a case of color (red or white) by hearing! The men who oppose evolution may be generally divided into two classes,—those actuated by theological prejudice, and those who know nothing about it.

(2) The theory of evolution is constructed out of the observed and accumulated facts of the universe: it is not guess-work. The men who have elaborated this answer to the old question, How did things come to be as they are? are men who have gone to the facts themselves, and asked the question. They went to the earth and studied it, and so developed the science of geology: they looked at the stars to see how they moved, and so made astronomy: they studied animals to see how they grew, and so made zoology: they studied man, and so made physiology and anthropology. If anybody, then, in the world, has

any right to an opinion on the subject, it is those who
have looked at the facts to find out about them. And
it is simply absurd to see people offer an opinion, who
have no better stuff than ignorance or prejudice to make
it out of.

(3) It stands the very highest test of a good theory;
that is, it takes into itself, accounts for, and adjusts, almost every known fact; while there is not one single fact
known that makes it unreasonable for a man to be an
evolutionist.

Now, what is the theory? Simply this: that the
whole universe, suns, planets, moons, our earth, and every
form of life upon it, vegetable and animal, up to man,
together with all our civilization, has developed from a
primitive fire-mist or nebulæ that once filled all the space
now occupied by the worlds; and that this development
has been according to laws and methods and forces still
active, and working about us to-day. It calls in no unknown agency. It does not offer to explain a natural fact
by a miracle which only deepens the mystery it attempts to
solve. It says, "I accept and ask for only the forces that
are going on right before my eyes, and with these I will
explain the visible universe." Certainly a magnificent
pretension, and, if accomplished, a magnificent achievement, of the mind of man.

Look at the theory a little more in detail. Evolution
teaches that the space now occupied by suns and planets
was once filled with a fire-mist, or flaming gas. This
mist, or gas, by the process of cooling and condensation,

and in accordance with the laws of motion naturally set up in it, in the course of ages was solidified into the stars and worlds, taking on gradually their present motions, shapes, and conditions. This is the famous "nebular hypothesis." In favor of this theory is the fact that the earth is now in precisely the condition we should expect it to be, on this supposition. The moon, being smaller than the earth, has now become cold and dead. Jupiter and Saturn, being larger, are still hot, — halfway between the sun's flaming condition and the earth's habitable one. And then all through the sky are clouds of nebulæ, still in the condition of flaming gas, whirling, and assuming just such shapes as the evolution theory alone can explain. The theory further teaches, that, when the cooling earth had come into such a condition that there were land and water and an atmosphere, then life appeared. But how? By any special act of creation? No. It introduces no new or unknown force, and calls for no miracle. Science discovers no impassable gulf between what we ignorantly call dead matter, and that which is alive. It does not believe any matter is dead: so it finds in it "the promise and potency of every form of life." It has discovered a little viscous globule, or cell, made up chiefly of nitrogen and albumen. It is a chemical compound, the coming into existence of which is no more wonderful than the formation of a crystal, and calls no more urgently for a miracle than a crystal does. This little mass, or cell, is not only the lowest and most original form of life, but it is the basis of every form. There is no single form of life on the globe, from the moss on a stone up to the brain of Sir Isaac Newton, that is not a more or less complex compound or combination of this primary, tiny cell; and there is no stage in the process of development, where ascertained laws and forces are not competent to produce the results. There is no barrier between the vegetable and animal kingdom. No naturalist living can tell where the one leaves off, and the other begins, so insensibly do they merge into each other, like day passing through twilight into night. Neither is there any barrier between species, either of plants or animals. This point is now settled. Evolution also (what no other theory does) explains the distribution of plants and animals over the surface of the earth. It explains the present condition of the races of mankind, — the progress of some, the stagnation of others, and the cases of gradual decay and dying-out. It explains social, political, and religious movements and changes, rises and falls. It is gradually proving its capacity to grapple with and solve the great enigmas and questions of the ages. And when generally understood and accepted, it will modify and direct all the forces and movements of the modern world.

From the primeval fire-mist, then, until to-day, the world has grown, without any necessity for, or help from, special creations, miracles, or any other forces than those known and recognized as at work right around us. It has taken millions of years to do this; but what are they in eternity? There have been no cataclysms, nor breaks, nor leaps. The sun has shone, the rain has fallen, the winds have blown, the rivers have run, the oceans have worn the shores, the continents have risen and sunk, just as they are doing now ; and all these things have come to pass.

But some one will say, "This is blank and outright atheism. You have left God entirely out of the question. Where has he been, and what has he been doing, all these millions of years? From the fire-mist until to-day, all has gone along on purely natural principles, and by natural laws, you say?" Yes, that is just what evolution says. But, before we call it atheism, let me ask you a question. Here is a century-old oak-tree. The acorn from which it sprung was the natural product of some other oak. It fell to the earth, and the young oak sprouted. From that day to this, — a hundred years, — the oak has simply grown by natural law. You want no miracle to explain it. Is your theory of the oak, then, atheistic? Is it any less strange that the oak should grow than that thousands of other oaks, and other forms of life, should do the same? When a child is born, it grows, you say, by natural law. Is it any more wonderful that it should be born by natural law? and that all life should be born, and should develop, by natural law? You are just as atheistic to say that a tree or a child grows by natural law, as evolution is, when it says the world did the same. Suppose science should put its God back in the past some millions of ages, while Moses puts his back only six thousand years, would the difference in time make one theory more atheistic than the other? But I should call pushing him back six thousand years, or a hundred million years, or five minutes, even, more atheistic than I should like to believe. So I would do neither the one nor the other. What if we see the life and power and movement of God in the fire-mist, in all the growing worlds, in the first appearance of life on the planet, in the forms that climb up through all grades to man? What if we see him in the dust of the street, in the grasses and flowers, in the clouds and the light, in the ocean and the storms, in the trees and the birds, in the animal, lifting up through countless forms to humanity? What if we see him in the family, in society, in the state, in all religions, up to the highest outflowerings of Christianity? What if we see him in art, literature, and science? What if we make the whole world his temple, and all life a worship? All this we may not only do in evolution, but evolution helps us do it. I shall be greatly mistaken, if the radicalism of evolution does not prove to be the grandest of all conservatism in society and politics not only, but in religion as well. It will turn out to be the most theistic of all theisms. It will give us the grandest conception of God that the world has ever known. It is inconsistent with "orthodoxy," but not with religion. It is charged by the thoughtless with being materialistic; but in reality it is any thing else. It so changes our conception of matter as utterly to destroy the old "materialism." It not only does not touch any one of the essential elements of true religion, but, on the other hand, it gives a firm and broad foundation on which to establish it beyond the possibility of overthrow. To illustrate this will be the work of future treatment of the special topics.

It only remains for me now to suggest a comparison as to grandeur and divinity between the two theories of creation. So many thoughtless sneers have been flung at the theory that dared talk of man's relationship to the ape, that a comparison like this may help change the sneer to admiration.

We marvel at Watt, the first constructor of a steam-engine; but it has taken many a brain beside his to bring it to its present perfection. What if he had been able to build it on such a plan, and put into it such a generative force, that it should go on, through long intervals of time, developing from itself improvements on itself, until it had become adapted to all the needs of man? It should fit itself for rails; it should grow into adaptation for country roads and city streets; it should swim the water, and fly the air; it should shape itself to all elements and uses that could make it available for the service of man. Suppose that all this should develop from the first simple engine that Watt constructed; and should do it by virtue of power that Watt himself implanted in it? The simple thought of such a mechanism makes us feel how superhuman it would be, and how worthy of divinity. Is it not infinitely more than the separate construction of each separate improvement? And yet this supposition is simplicity and ease itself, compared with the grand magnificence of creation after the Darwinian idea. Who can pick an acorn from the ground, and, looking up to the tree from which it has fallen, try to conceive all the grand and century-grown beauty and power of the oak as contained in the tiny cone in his hand, and not feel overwhelmed by the might and the mystery of the works of God? How unutterably grander is the thought that the world-wide banyan-tree of life, with all its million-times-multiplied variety of form and function, and beauty and power, standing with its roots in the dust, and with its top " commercing with the skies," and bearing on its upper boughs the eternal light of God's spiritual glory, is all the godlike growth of one little seed in which the divine finger planted such fructifying force!

III.

THE GOD OF EVOLUTION.

THE manifestation of the life and power of the universe has been a gradual evolution; that is, a continuous unfolding, a growing revelation, the less becoming more, the simple becoming complex, through an apparently infinite development and specialization. The whole is recapitulated in every flower that blooms. The rose is first a seed; it pushes up through the soil, then branches, sends out bud and leaf, bursts into beauty and fragrance of flower, and then develops from its loveliness the germ of something more to come. So of the mountain-pine, that has stood on its watch-tower overlooking whole centuries of human history. Its potential life was once wrapped up in one little seed, sheltered in the protecting grasp of a tiny cone that one might hold in his hand. The might and the marvel of this life are not to be seen by simply looking at seed or cone, but by studying the processes and results of the wondrous unfolding, or evolution.

The universe, so far as at present we are able to trace, was once contained in the world-seed of the fire-mist, or primitive nebula. We might guess all sorts of things, as to where the fire-mist came from, and how it came: but science means knowledge; and it is my purpose, so far as possible, to keep myself strictly to what is known. Theology assumes God, and thinks thus to escape all difficulties. But should science take to assuming, it has the same right to declare that the fire-mist made itself, or that it always existed, as theology has to say that God made himself, or always existed. I am willing frankly to admit that it is just as easy to think that the universe always existed, or that it made itself, as it is to think that God always existed, or made himself. It is simply impossible to imagine or comprehend either the one or the other. As, then, I wish to speak to and command a respectful hearing from those who are not content to rest in tradition or assumption, but who wish to find out if religion has any basis of knowledge to stand on, I propose to avoid assuming any thing, even the existence of God. I wish to find him, if I can; and then there will be no need of assumption.

We begin, then, with the fire-mist, because that is the first thing we know. This fire-mist, by cooling and condensing, became suns and planets and worlds, — the wondrous heavens of our present telescopic astronomy. Our system was slowly born. The sun, so large that it has not yet cooled, is the source of our life and light and heat. The moon, so small that it is already cold and dead, exists now only for the sake of the earth. Other members of the system are still half-sun and half-planet, not yet ready for the abode of life. When it was ready, the lowest forms of life, having nothing of limbs, or bone, or brain, — no organization, — appeared in the primeval oceans of the world. Then there were long ages, during which the highest type of life on the earth were fishes. The whales and leviathans were the kings and the nobility of the time. Then these climbed up into the higher form of reptiles. Huge creatures, half living in the seas, half on the land, or with great dragon-wings flying heavily through the air, now reigned over the earth for thousands of years. Then came the birds; and for centuries they were the highest form of life upon the globe. Next animal shapes appeared, climbing up from the lowest types and simplest forms, until the great tree of life flowered out into humanity. But the first being that could be properly called human was, as compared with later developments, no more than the insignificant blossom of a wayside weed when placed beside the perfumed glory of the rose, or the gorgeous tropical outflowering of the century-plant. The gulf that separates the highest animals from the lowest men is as nothing compared with the wider differences that divide between those lowest men and the Dantes, the Shaksperes, and the Newtons of the race. And above these the moral grandeur of one like Jesus towers like a mountain, that, above all its range, looks down on its fellows from the clouds.

Thus the panorama of creation has unfolded. The first scene was the fire-mist: the last that we have looked on is the present hour, including the highest social, political, and religious life and aspirations of the race. And the end is not yet: the creation is not still: the scene is moving and unfolding to-day. And if we may guess the future from the past, the imagination must confess that it has no colors bright and grand enough to paint the possibilities of the ages to come.

You can now see that the manifestation of the life and force of the universe has been a gradual evolution. And if, at any particular stage of the progress, there had been some person present trying to find out what the life was by looking at "the things made," of course he would have had to make his idea of this life, or God (if he gave it that name), out of the then condition of things. His god would have been on the level of the fire-mist or the fishes, like the Philistines' Dagon; or of the reptiles, like the deities of the serpent-worshippers; or of the animals, like the sacred bull of Egypt; or simply a physical man, like the Roman Hercules or Jupiter, or the early Hebrew Jehovah (for Jehovah at first was only a gigantic man): as David says, "Jehovah is a man of war." As such he appears all through Genesis.

This manifestation of the life of things has thus been a gradual and growing one. And it is curious and instructive to notice that the idea of God in the human mind has recapitulated, or lived over again, this evolution of the facts of the world. Have you ever thought over the first conceptions that men had of God, and how they have developed to what we think and believe to-day? The first men of the world, of course, had no fire, no houses, and no weapons with which to defend their lives, or to hunt for food. The most important invention of the world was discovered when some man first made a spear out of a stick. Defenceless as they were in the midst of hunger and cold and storm and wild beasts, the predominant motive and feeling of the time must have been fear; and from this would naturally develop their first religion. As they saw each other and the wild animals moving about, and crying, and making noises, and saw that they were alive, they, like the child-men they were, thought that every thing that moved, or made a noise, was also alive. And as the wild beasts hurt and killed them, when they got a chance, and as the storms and cold hurt and killed them, their most pressing thought was one of safety. They saw the trees move when the wind blew; they felt the motion and power of the wind that they could not see; they saw the waters run; they saw the lightning flash, and twist itself like a huge serpent in the clouds; they saw the clouds go across the sky, and the sun and moon rise and set; and as they saw all these things moving in this way, they thought they were living beings, who might hurt them if they were angry, or might help them, if they chose. Thus they turned all these things into gods. Then they began to pray to them, and to give, or offer, them things they thought they would like, and to try to find out how they might influence them to do what they desired. This was the original polytheism, or, in its lowest manifestation, fetichism.

But after a good many ages, some of the races of men

grew wiser. They found out that stones and trees and rivers and clouds and the sun were not gods. They learned that praying to them did no good. They discovered that, instead of moving about as they liked, they were governed by some higher power, and moved according to certain definite laws. Should a barbarian look at a complicated piece of machinery, he would be likely to think, at first, that every several wheel and lever and piston and band was going all by itself. But study would teach him that a central force moved it all according to the laws of its construction. Something like this men found out about the world. They learned that there was somewhere a hidden mainspring that controlled the whole life and movement of the universe, and so gave unity to it all. This was monotheism.

But, as I have said, the first development of monotheism only thought of God as a gigantic and superhuman man. The Jehovah of the Old Testament makes himself a local habitation, appears in the temple, walks and talks, and thinks and plans, loves and hates, gets angry, takes vengeance, and changes his mind, very much after the fashion of an Oriental despot. This is not to be wondered at; for as water cannot rise above its source, so the human mind cannot think of God as being any thing higher than its highest and best conception of what is worthy of divinity. Humanity cannot escape itself; and so its thought of God is always the best it is capable of thinking at the time. As man grows and develops, so does his idea of divinity. The divine does not change;

but as you can put only twelve quarts of the Atlantic in a twelve-quart pail, so in a finite brain you can put only so much of the Infinite as the finite can contain. As the thought of man gets larger, its contents increase.

The next grand step in the development of monotheism is when Jesus says, "God is a spirit," "The hour cometh when ye shall neither in this mountain [Gerizim], nor yet at Jerusalem, worship the Father." Thus, to the highest thought of men, God has become a power, whose centre is everywhere, and his circumference nowhere.

This developing idea of God is a fact of human experience. But now the question meets us, Is this idea any thing more than an idea? Has it any real validity, or right to exist? Do we know that there is any God corresponding to the idea? Knowledge is popularly supposed to have for its province all the regions of the tangible and definite; while the indefinite, and the intangible, and the unseen, are turned over to faith and imagination. And a great many hard-headed people are beginning to think that what they cannot see and feel is of no practical consequence. Even Tennyson, speaking of religion, permits himself to write: —

> "We have but faith : we cannot know;
> For knowledge is of things we see."

We are accustomed to say, "I know the earth. I know a crystal. I know my friend." But when we come to matters of religion, we say, "I hope, or I believe, that God exists."

Now, I protest against this use of language, as not being true to the reality of things. I know it will be regarded at first, and by many, as daring and presumptuous; but I propose to make and substantiate the claim that the God of evolution, the hidden life and secret force of this unfolding universe of ours, may be just as truly and really known as a grass-blade, a star, or your next-door neighbor. Please take notice that I do not say as well known, as completely known, but known as truly and really, as far as the knowledge goes. There is nothing in the universe that is completely known. For the universe, like the seamless garment of Jesus, is all of one piece: every thread runs through it all; so that to trace one thread completely is to unravel the whole mystery. As Tennyson says with such force and beauty, —

> "Flower in the crannied wall,
> I pluck you out of the crannies,
> Hold you here, root and all, in my hand,
> Little flower; but if I could understand
> What you are, root and all, and all in all,
> I should know what God and man is."

To justify this statement, I must ask you to consider, for a moment, what it is that you know about any thing. Is it the thing or the person in itself, or manifestations of the thing or person? In what sense do you know your next-door neighbor? You know the general size and features of his person, the color of his hair and eyes, his gait and style of movement, the clothes he ordinarily wears. You know the tone of his voice, something of his past experience and present character, and, in some general way, his intellectual and moral attainments; that is, you know certain external manifestations of this mysterious personality you call neighbor and friend. And no matter how intimate you may be, the general fact is not changed. In the secret recesses of his life, there is a whole unvisited world. There are oceans on which you have never sailed, continents you have never looked upon, and skies whose stars have never caught your eye. In short, you know nothing about the essential personality of this mysterious being you call friend. You simply know certain outer manifestations of the inner life. And the same is true of a flower, of a piano, or of a chair, just as really as it is of the sun, or of God. All these have certain qualities that manifest themselves to your senses. You have five senses to receive the manifestations. But that your five senses and these manifestations are all there is, is the boldest of all assumptions. As reasonably might a deaf and blind mute assert that there was nothing in the world but hardness and shape, because these were all he could feel.

"There are more things in heaven and earth, Horatio,
Than are dreamt of in your philosophy."

We have a fairly accurate knowledge of manifestations. It is not perfect; for our senses sometimes make mistakes. But one thing more certain than any other item of our knowledge is, that beneath all manifestations is a life and a force of which the manifestations are the outcome. I may mistake as to some special fact about my friend; but that there is a personal, conscious life that I call my friend is matter of absolute certainty. I may be color-blind, and so mistake the red or the purple of a rose; but that there is something that manifests itself to me as a rose, this I absolutely know. And so the one thing that we know about the universe more certainly than we know any thing else is, that, underneath all the forms and movements and manifestations, there is a life, or a force, that manifests itself as form and movement. And as our knowledge of a grass-blade or a friend is only a series of facts of manifestation, so we have concerning God precisely the same means of knowledge. Let us go on, then, to indicate the outlines of what we can know about the life and force of the universe that theology calls God, and that science may call "nature" or "law," if it chooses, since the name, whatever it be, cannot change the essential reality.

(1) We have that whole class of manifestations, which, taken together, make up the material universe. And what can we spell out of these wondrous hieroglyphs? They certify, at the outset, that this which religion calls God, and which science names force, or power, or nature, really exists. The universe is, and therefore that which the universe manifests is. Next, here is everywhere the movement of life; and beyond motion there is order, — an order that speaks of a cosmos, or system. I do not care to assert that this order demonstrates a designing mind. It is sufficient for my purpose, at this stage of my argument, to say that this order is such an one as agrees

with the highest development of what we call intellect; so that, looking up to it from our human plane, it is proper for us to call it an intellectual order. And, further still, we see everywhere, on leaf and flower, on sunlit cloud, on curl of ocean surf, in mountain outline, and in wildwood glade, an inexpressible beauty that becomes the inspiration of artist and of poet.

Here, then, in inanimate nature, we see and know existence, motion, order, and beauty, and know them as the outcome, as the real and true manifestation, of the inscrutable and ineffable life of the universe. So far as they go, they are reliable and adequate revelations of the Unknowable One.

(2) As our next step up and on, let us glance at the lesson of human history. I do not here enter into any analysis of human nature, but wish simply to ask your attention to the way in which the universe has dealt with the race, taken in the mass. What is the lesson of the drift of our human destiny, taking the world as a whole? It is twofold, and may be indicated by the two words, "progress" and "righteousness." From the lowest forms of primeval life up to the topmost height of our modern civilization, there is evident a force of uplifting and onlooking.

> "Every clod feels a stir of might,
>   An instinct within it that reaches and towers,
> And, groping blindly above it for light,
>   Climbs to a soul in grass and flowers."

What the poet here sings of the lower life of the

spring may be taken as typical of the grand truth that binds together the Alpha and the Omega of creation.

And not only do we see progress along certain definite lines of law that suggest the rightness of this life-force of the universe, but this progress has lifted up into what we call the sphere of morals, and has been along certain other definite lines of what we call righteousness; so that the lesson which Matthew Arnold so finely deduces from the history of Israel may be read with more emphasis still in the history of the race. This power of progress is also a power of progress toward a moral ideal, — "a power that makes for righteousness." I need hardly illustrate this; for I suppose no one but a pessimist will hesitate to accept it. A full illustration would be an outline of universal history. This power making for righteousness has been that by which nations have grown, or the rock on which they have foundered. The nations which to-day stand highest in civilization are those which, on the whole, best conform to and live out this law.

And, indeed, not only is this so, but it is easy to see that it *must* be so. For as, in the body, if once disease gains supremacy over the healthy powers, death must ensue, so in the universe, if the lawless and evil forces were really in the majority, the cosmos would tumble into chaos.

This inscrutable power of the universe, then, is progress and righteousness as manifested in the outlines and drift of human history.

(3) Come, now, to man, individual and social, and see

what is manifested here. And just at this point I must
stop to notice, and protest against, that which seems to me
the height of what is irrational and strange in the reasoning of many, on the side of both science and theology.
Theology, in its attempt to exalt man, takes him out of,
and sets him apart from, the order of nature, and then
abuses nature as an untrustworthy guide in religious
things, because it does not find moral qualities — love
and mercy — in stones and mountains and trees. It
takes the soul out, and then wonders that it does not have
any soul. And many scientists, as if willing to take
theology at its word, go ranging through the inanimate
universe, as though they were examining some mechanism
with which they had nothing to do, and declare that they
do not find what nobody supposed they would find in the
material forms of the world. As if the mainspring of a
watch should start into independent life, and go to search-
ing through the rest of the machinery in the attempt to
find that of which itself was the representative, and
should then declare, on its honor as a good piece of steel,
that the watch showed no signs of a mainspring, and thus
was radically defective!

Whether or not there be any thing about man rightly
called supernatural, we know, that, at any rate, he is
natural. He is a part of the life and order of the world;
and thus, in all the myriad manifestations of his varied
life, he is an outcome of the central power and life of the
universe. He is a part of the divine manifestation.
What, then, are the things that he reveals?

In him, first, so far as we know, does the world come up into a consciousness of personal existence. While this does not prove that the inscrutable power of the world is a person, it does prove that this power is, at least, as much as, and as good as, personal.

Then, beyond this personality that consciousness reveals, man manifests all those qualities that we call social, moral, and spiritual,—love, devotion, self-sacrifice, purity, integrity, patriotism, heroism, and the "enthusiasm of humanity." The mother watches tirelessly over some sick child; or she gives herself to the care of one idiotic or deformed. She forgets selfishness, and finds pleasure in wasting away, and wearing out her life, for the sake of her mother-love. Men ride at the front of embattled armies, meeting danger, nor shunning death, for the triumph of some noble sentiment or intangible principle. Winkelried takes the spears of the foe into his own breast, that he may make a breach through which his followers may pour to a patriotic victory. Mattie Stephenson goes to plague-stricken Memphis, and gives her life to a pure pity for the helpless sufferers. John Brown "counts not his life dear," so he may be able to help a degraded mass of slaves up into freedom and manhood. The religious martyr stands chained to the stake, with the kindling fagots about his shrivelling limbs, when one false word would set him free; and when the last flame leaps up, his life goes out on the air that still trembles with his song of triumph—and all for what he holds as sacred truth and divine light. A Jesus or a Socrates dies peaceful and calm, while with his last breath he forgives, or prays for, the ignorant rage, or pitiable malice, that puts him to death.

Now, this whole realm of the moral and spiritual is a real country. These things are facts of human life and history, — just as much facts as the labelled fossil bones and the flint arrow-heads of scientific museums. These facts are real and verifiable manifestations of the power of which all the phenomena of the universe are expressions. And if this power cannot be adequately expressed in these terms of humanity, it is at least certain that it is as much and as good as these. The partial expression is not false: it is only inadequate. This power, then, is as good and loving, and pitiful and devoted, as the best manifestations of itself in humanity.

(4) One step more we will take. Above the common level of our humanity there rise the exceptional and towering summits of those mountainous men — seers, prophets, poets, lawgivers, leaders of every kind — that have served as landmarks and observatories for the race.

Consider, for a moment, the significance of the fact that there have been such men. If a daisy springs out of the sod, it is because there was a daisy in the sod. If such men spring out of humanity, it is because there is in humanity the stuff of which such men are made. If a man admires the grand and sublime, it is because there is grandeur and sublimity in him to respond to the outer appeal. And thus the fact that the race is seen on its face, adoring the idealized forms of these sublime and divine men, is proof that humanity is potentially such as they.

Now, these seers and great ones of the world have manifested something still more and higher than the common life of the race. They have seemed to be conscious of the divine and eternal as other men are conscious of themselves. Above the low-hanging clouds, like mountain-peaks that forever look in the face of the clear heaven, and gaze on the unsetting stars, they have looked on the face of the divine, and have been conscious of fellowship with it. And they have so dwelt in the unchanging and permanent principles of life and truth, that they have felt that they tasted immortality. I claim for these grand visions and hopes only what they were, — grand visions and hopes. But, as such, they were facts, had a source and meaning; for the vision of an ideal can no more come out of nothing than can a mountain or a world.

Humanity, then, at its summit, has had these outlooks, gained these glimpses of something better than it ever saw realized, and gazed at a "light that never was on sea or land."

What, now, has been so far developed? The material world is a manifestation of existence, order, beauty, and power. History, in the physical world and in humanity, is the manifestation of progress. Human history tells of a "power that makes for righteousness." The ordinary life of humanity speaks of love, devotion, hope, self-sacrifice, purity, pity, and all the range of powers and faculties that we call moral and spiritual. The seers have had visions, and have dreamed dreams, of a divine life with which they could commune, and of an immortality they could consciously taste.

And these things all are facts of human knowledge. And they all are outcomes and manifestations of the infinite and inscrutable life and power of which all phenomena are expressions. So much, then, we know about God. The claim is not made that they are adequate, or that God is any one of these things, or all of them put together. But the claim is made, that these are manifestations of God just as really and truly as color and odor are manifestations of a lily; and though we cannot say, "These are God," or "God is one or all of these," we can say that "He is as much as and as good as these not only, but that he fills all these conceptions, and overruns them infinitely on every side, just as the light fills and overflows the goblet of the sun." God is, then; and he is infinitely beyond any conception we can frame, or any manifestation we have seen; and Paul was right when he wrote, "The invisible things of him from the creation of the world are clearly seen, being understood by the things that are made."

From "the things that are made," then, and as matter of pure science, we know that God is. And we know that in him are qualities that manifest themselves as power, order, intellect, beauty, love, pity, devotion, and all that we call by the terms "moral" and "spiritual." You may call them attributes; you may call them by some other term; you may not call them at all: but the fact that these qualities are manifested remains just the same.

And now I must call attention to two or three ancient, still unanswered, and perhaps I may safely say unanswerable, objections that lie against the nature and character of God as conceived by the old creation theory of the world.

(1) If he is objective to and outside of the universe, as the architect is outside of the house that he builds, then he is not infinite, and so falls short of our necessary idea of Deity. Men speak of nature as separate from God, and of God as making and ruling nature, as though it were a sort of machine which he constructed and runs. But the infinite must include the all. If there is any real entity that is not divine, then God is not king. If he be not in the dust of our streets, the bricks of our houses, the beat of our hearts, then he is nowhere. The old theory destroys the infinite, and only gives us two finites in its place. This is a relic of the old dualistic belief that modern thought is swallowing up in its all-inclusive unity. If no other objection could ever be brought, this alone would be sufficient to make the common idea intellectually untenable. This difficulty is perfectly met and answered by the evolution theory.

(2) If God consults and thinks and plans, as Genesis represents him doing, then he is simply a magnified man made in our image. This kind of deity is only the spectre of the Brocken, the gigantic shadow of man himself projected against the clouds. Against this might be brought

the old prophet's accusation, where he represents God as saying, "Thou thoughtest that I was altogether such an one as thyself; but I will reprove thee."

(3) Then, again, on the old theory, it is impossible to relieve God of the responsibility of the authorship of evil. Brave old Lyman Beecher may say, "I don't want any son of mine to defend the character of God. He is able to look after himself." But the popular God of Christianity needs the defence, nevertheless. And all the theological works on the divine government ever written have not defended him, either. Milton started out to "justify the ways of God to man," but succeeded only in leaving it a question as to whether Satan were not the hero of his poem. Edward Beecher postulated pre-existence as the answer; but this only made the charge more ancient, without clearing it up. There is no use in trying to evade it. An all-wise, all-powerful, and all-loving God might have made a better world. And to say that he made it all — good and bad, and both eternal — "for his own glory" is only to add infinite selfishness and egotism to the original difficulty; also that, being God, he had a right to do as he chose, is only to justify the Neros and Napoleons of history by making divine the infamous cruelty that "might makes right." The world is neither physically nor morally perfect; and John Stuart Mill only voices the thought of all earnest and honest minds, when he praises his wife as one "free from that superstition that ascribes a pretended perfection to the universe." He says, if God made the best world he could, then he is

not almighty; and if he could have made a better, and did not, then he is not perfect goodness. If the Church says, "The earth is cursed for the fall of man," then who made man fall? If it answer, "The devil," then who made the devil? and if it still say, "Either devil, or man, or both, chose freely to sin," then why did God permit it? or why make them so that they would fall? The difficulty only shifts and changes: it is not removed. It always comes back on God, after all.

But evolution at least hints a satisfactory reply. By this theory, the universe is still growing; and by the very terms of the conception, any thing that is growing is never, at any particular stage of the growth, complete. You would not criticise a picture still on the easel: wait till the artist is done. The summer-sweeting is bitter the first of June. The old hymn, then, is good science:—

"His purposes will ripen fast (slow),
Unfolding every hour;
The bud may have a bitter taste,
But sweet will be the flower."

And then the drift of science is pointing toward the probability that evil is not a real thing at all. It is only temporary maladjustment, a condition to be outgrown. Evolution, then, may believe in a perfect God. It asks not unreasoning faith. It plants itself on the solid facts of the past, and waits to see the same forces unfold ever new and newer glories, and justify grander and still grander hopes.

It remains to notice one or two questions or objections that must be disposed of before our subject is complete.

(1) The materialist comes and says, "I grant that all these things that you have said are facts; but how do you know that every thing is not the result of the various combinations of matter? Who shall say that the poems of Shakspere, the science of Newton, and the religion of Jesus, are not the outflowering of the higher and finer forms of matter, just as much as the rose is a blossoming of the dust?" Such questions as these are asked by some of the best thinkers of the time. We cannot afford to slight them.

I have two suggestions by way of reply. And, first, if you say these things that we call intellectual and spiritual inhere in and come out of matter, then you change the whole conception of the term, and matter becomes something spiritual and divine. This is utter destruction of the old materialism, and makes matter only a form of the eternal. It simply converts matter into what we mean when we say spirit. But, in the second place, I challenge any man living to prove that matter is a real and substantial existence in itself, and as separated from the force and life that we call spirit. So far as any man can tell, matter is only the robe that spirit and life eternally weave for themselves, and no more capable of separation from them than light can be separated from sunshine. What matter is, or what spirit is, in itself, nobody knows: so that to say there is nothing but matter, is just as offensive and unwarranted dogmatism as is any ecclesiastical claim whatever. Even a man like Huxley will say that the idealism of Berkeley is more consonant with our present knowledge than is dogmatic materialism.

(2) Another question: Is the God of evolution a personal God?

Here, again, I must make a twofold answer. If your term "person" implies what we mean when we say Gen. Grant is a person, or Queen Victoria is a person, then any thoughtful mind will have to say No: God is not a person in that sense. This kind of personality is limited, outlined, localized. Any true thought of God recognizes him as infinite; and the infinite cannot be bounded, outlined, or localized. You must not paint God as Moses and Michael Angelo did, as only a great man, exalted, and sitting on a throne, even if you give him the brow of Jove, and put the lightning in his grasp. Jesus asserted the higher idea when he said, "God is a spirit." He is not on Zion or Gerizim alone, but everywhere. But, on the other hand, if by denying him personality you take away from him something, and lower him in your thought, you must not do it. He is not less than personal, but infinitely more. Personality, in man, is one of his minor manifestations; and that which is manifested is not something more, but something less, than that which manifests. Nothing comes out of nothing; neither does the greater come out of the less. God, then, is all of good and great and helpful that we mean by personal; and beyond that he is infinitely more than the word "personal" can express.

Who, then, is the God of evolution? Not the mechanical contriver, or the Oriental despot of the Old Testament; not the Zoroastrian Ahura-Mazoa, ruling but half the world; not the Hindoo Brahm asleep in the heavens; not a deity dwelling in temples, and only to be sought at special altars; not the partial and implacable God of Calvin; not one sitting afar on his throne, to be reached only through mediators. The righteousness which is by evolution speaketh on this wise: Say not in thy heart, Who shall ascend into heaven to bring him down? nor, Who will descend into the deep to bring him up? But what saith it? God is nigh thee, even in thy mouth and in thy heart. And it says this with a reality and meaning never said before. Or it borrows the beautiful and mystic tongue of Wordsworth, and speaks of

> "A sense sublime
> Of something far more deeply interfused,
> Whose dwelling is the light of setting suns,
> And the round ocean, and the living air,
> And the blue sky, and in the mind of man;
> A motion and a spirit that impels
> All thinking things, all objects of all thought,
> And rolls through all things."

Or, with Alexander Pope, it is ready to run its faith into music, and sing, —

> "All are but parts of one stupendous whole,
> Whose body nature is, and God the soul;
> That, changed through all, is yet in all the same;
> Great in the earth as in the ethereal frame:

Warms in the sun, refreshes in the breeze,
Glows in the stars, and blossoms in the trees,
Lives through all life, extends through all extent,
Spreads undivided, operates unspent,
Breathes in our soul, informs our mortal part,
As full, as perfect, in a hair as heart;
As full, as perfect, in vile man that mourns
As the rapt seraph that adores and burns;
To him no high, no low, no great, no small:
He fills, he bounds, connects, and equals all."

IV.

THE MAN OF EVOLUTION.

One of the most noted sayings of the early philosophy in Greece was contained in the two words, "Know thyself." And, however much we may be interested in stars or earth or animals, yet history, biography, epic poetry, and the universal love for novels, tragedy, comedy, and stories, show that to man the most interesting thing in the world is humanity. Even trivial gossip is only interest in our fellow-creatures that has turned a little sour. Thus the nature of man, his origin, and how he came into his present condition, and the drift of his true progress — these are the most practical of all questions. And all the great concerns of the day, — religious theory and experience, matters of reform, how to deal with crime, methods of politics and government, — all must find their ultimate solution in the nature of man. The farmer, the physician, the chemist, the carpenter, the worker in metals, all practical laborers, know that their success depends upon their knowledge of the materials in which they work. The stonecutter cannot hammer his blocks into shape any more than the blacksmith can grind or chisel his. The work all turns on an accurate knowledge of the material. So it is beginning to be found out that much of the religious, philanthropic, and political work of the world has been thrown away for lack of a true knowledge of the nature, the capabilities, and the needs of humanity.

But to what source shall we go to learn the nature of man? For ages men took all their ideas about the stars and the earth and the animals from certain ancient records of what the men of old time thought about these things. But at last it occurred to them to study the stars and the earth and the animals; and from that study they learned that all those ideas were wrong. It took the world a long while to learn that the best way to find out about them was to look at them. A few people are just beginning to wake up to the notion that the best way to learn the nature of man is to look at him. In the words of Oliver Wendell Holmes, "We must study man as we have studied the stars and the rocks. We need not go to our sacred books for astronomy or geology. Do not stop there. Say now bravely, as you will sooner or later have to say, that we need not go to any ancient records for our anthropology. Do we not all hope, at least, that the doctrine of man's being a blighted abortion, a miserable disappointment to his Creator, and hostile and hateful to him from his birth, may give way to the belief that he is the latest terrestrial manifestation of an ever upward-striving movement of divine power? If there lives a man who does not want to disbelieve the popular notions about the condition and destiny of the bulk of his race, I should like to have him look me in the face and tell me so." And he adds, "We have taken the disease of thinking in the natural way. It is an epidemic in these times; and those who are afraid of it must shut themselves up close, or they will catch it."

It is time, then, that we studied man to find out about man; he is not inaccessible, a great way off, and hard to come at: he is the nearest fact of life. We can look at him, and find out about him. And within the last fifty years the records of his genealogy have been discovered. We can now, in spite of some gaps in it, trace the line of his descent, and find out where he came from, and by what steps he has progressed.

There are two great and fundamentally opposite notions concerning human nature, that, with sufficient accuracy, I may call the Oriental and the Occidental, the Eastern and the Western. The Chinese, the Hindoo, the Arabic, the Jewish — these may represent the Oriental. The Greek, the Roman, the German, the English, and the American, may represent the Occidental. The Occidental is the theory of self-respect. All our modern civilization is the result of it. It believes in the grand capacity and noble possibilities of the race. It seeks to make the most of itself, and of the exhaustless resources of the world. The Oriental is the theory of self-contempt. It casts dust and ashes upon its head, and lays its mouth in the dust before some supposed divine despot. It looks on life as mean, and the body as vile. It is a part of the belief that all matter and all life are evil. The Brahmin and Buddhist aspiration to escape the curse of life by absorption in deity, or the calm of practical annihilation, is its natural fruit in religion. Political stagnation, social degradation, and a listless submission to cruel and fickle despots, are the natural fruits in practical affairs. The absurdity of the mixture of the two may be seen in church on almost any Sunday, when some self-respecting, wealthy, and ambitious man, who is doing his utmost to get on in the world, mumbles over after the priest his Sunday creed — so different from his Monday one! — that life is a "vain show," wealth a snare, the ambitions and successes of life a delusion, and he himself a "miserable offender;" when, if anybody else should call him a "miserable offender," he would stand up in his dignified self-respect, and knock him down. It only means that he is living an Occidental life, and that he has inherited an Oriental creed.

The orthodox conception of man and his relation to God, total depravity, supernatural redemption, and eternal punishment, these are the outcome of the Oriental theory that is a part of our religious inheritance. In the East it took two forms; one, that being connected with matter at all, being born into fleshly bodies, was the source and cause of all evil. This is Hindoo, Buddhist, and Plato; and, through the Alexandrian schools of Philo and Neo-platonism, it has tainted and colored all our early Christian thought. The other is embodied in the story of Genesis. This allows that there was one man pure and holy in a natural body; but he early fell under the power of matter, and all his descendants are born immersed in it, and depraved.

Now, as this story of Genesis is the basis of the popular theology, and as this and the evolution theory are the only ones that earnest men in America are concerned with, we will narrow down our discussion to these.

The popular belief is that somewhere in the valley of Euphrates, God created, in the midst of the wilderness world, a garden of delights, a paradise of perfect innocence and beauty. In this garden, to dress and keep and enjoy it, were placed Adam and Eve. They two were physically and morally perfect. As old Dr. South expresses it, "An Aristotle was but the rubbish of an Adam, and Athens but the rudiments of Paradise." But right on the heels of perfection came utter ruin. A serpent — popularly supposed to be the devil, though the story says nothing of the kind, and the devil was not invented till ages afterward — is found equal to the task of frustrating the work of God, and seducing the obedience of man. There is little doubt that the serpent is only an allegorical figure setting forth the supposed sinfulness of fleshly desire, thus linking this story with the old notion that the flesh was inherently sinful. Any way, Adam fell; and God had created him in such unity of relationship with all his race that he dragged down with him all mankind. Heaven and hell hung on an apple-bough; and, when the fruit was tasted, —

"In Adam's fall
We sinned all."

The race was hopelessly lost. For "one man's disobedience" God, who just before had pronounced his work very good, turns all his love to hate. He curses the earth, even, for man's sake, and dooms him and his posterity to labor and sorrow.

Now, this is the corner-stone on which the whole system of orthodox theology rests. Total depravity, moral helplessness, infant-damnation, fore-ordination, limited vicarious atonement, and everlasting punishment — these all follow, with the fatal necessity of an irresistible logic, from the fall in Adam. And in the light of this necessary logic you may see the weakness of the positions occupied by the "liberal orthodoxy" of the time. The moral sentiment of the age has revolted against the dogma of infant-damnation, which was once universally held; and now it is cast to the rubbish heap of cruel superstitions. But he who believes in the fall of man must logically believe infant-damnation. And why it is any worse to damn an infant, than it is to damn a man who is born and who lives a helplessly and hopelessly depraved life of forty years, is difficult to see. The one was no more responsible for himself than the other. Then there are thousands who have rejected the fall, who yet cling to the atonement. But if man is not fallen, there is no need of an atonement. It must soon be seen that this whole system is an arch of doctrine in which every stone takes hold of, supports, and is supported by, every other stone. Knock out one of these, and the whole fabric tumbles into confusion; so that the only logical position is the old orthodoxy or reason. Such men as Dr. Gardiner Spring of New York did not fear to face and accept the logic, and say, "God saves and he damns just the number he wishes to in carrying out his own purposes."

Now, this doctrine, that God created man physically, intellectually, and morally complete, in a moment, and that by one sin he fell into a condition from which the present condition of humanity is the result, is surrounded by difficulties that seem to be insuperable. Let us glance at some of them.

And, in the first place, there is absolutely no proof of any thing of the kind. It is at the best an Old-World, Oriental tradition. There is not a single known fact that can be brought to its support, or that cannot be better explained in some other way. Were it not supposed to be divine revelation, no one would think of arguing in its favor. And the narrative is so mixed up with crudities and absurdities and contradictions and immoralities, that it is a dishonor to any high idea of God to suppose him capable of doing the deeds, or of inspiring a record of them.

In the next place, that God should make a man, or any thing else, full-grown in a minute, is utterly against all that we really know of the divine method. All things grow; and there is not a flower, or shrub, or tree, or animal on earth, that does not have on it, not only the marks of its individual development, but also the traces of an ancestral growth that reaches back into the earliest dawn of time. The thought is as absurd, and as incongruous with all our knowledge, as the boy's question when, in answer to his father's statement that God could do any thing he pleased, he asked him if God could "make a two-year-old colt in fifteen minutes."

Again, the story of Genesis does not account for the past history, the present abodes, and the different conditions, of the various races over the face of the earth. That a being, of whom Aristotle was only the "rubbish," should have produced descendants ranging all the way from Newton to the South-sea Islanders, the Patagonians, the Bushmen and Pygmies of Central Africa, is something that the fall of man does not explain. There is nothing in the nature of sin alone that can account for it. The influence of nature and the laws of life can account for all these varieties. And it is curious to notice, that, while orthodoxy denies that nature has any such power over the development and modification of man as evolution asserts, it is yet obliged to call in just the forces that science proclaims to explain the facts which it cannot deny. Nature asks as large powers to make all the races out of Adam, as she demands to make them out of protoplasm.

And then we know — it is no guesswork any longer: we know — that great civilizations existed, that men were born, loved, hated, hunted, fought, and died, thousands of years before Genesis says that man was created and fell.

And, once more, the theory (as I illustrated somewhat in the preceding paper) makes God a moral monster.

Whether, then, we can accept any other theory or not,

even if we have to go without a theory, this one we cannot hold. I might be content to say, "I know and can know nothing about God: I must walk my path of life in the dark, waiting to see what the future will develop;" but I cannot consent to say, "I love, I will worship, or even I respect, such a God as is taught by the popular theology." If there is a supreme power in the universe capable of making such a humanity as is preached, and of treating his child as he is represented as treating man, then though I may have to submit to his power, as feeble nations submit to despots, yet I will not, can not, love him, nor bow to him my knee. And if I must go to hell with the noble livers and the great thinkers of the world, then I would choose it rather than the place of court favorite in the presence of one who makes evil, and torture, and everlasting prison-houses, for his own glory.

We come now to consider the evolution theory of human nature, and to notice how far it corresponds with the known facts of the world. What is the theory? That all life on the globe is a unit, like a tree, and that man is the crowning blossom on the topmost bough. To confine ourselves to our immediate ancestry, it teaches that man has developed from the animal life beneath him. I am aware that the popular mind is full of prejudice on this subject, and that a large part of its impressions have been derived from newspaper jokes and caricatures. But I hope it will be remembered, that those persons who set up in the business of making people laugh are not particular about their materials. They will ridicule any thing that the popular taste will allow; just as Aristophanes wrote comedies to make Athens laugh at Socrates; and as the London "Punch" ridiculed and caricatured Lincoln because English opinion then favored it. I can sympathize with a man who shrinks from recognizing the ape as among his poor relations, particularly if there is a family likeness that he fears will be discovered; but really we must put prejudice one side: this is a matter not of feeling, but of fact. I, for one, am ready to think it far more wondrous and honorable that my body should have come to its present perfection by the marvellous pathway of the animal world, than that it came straight from the slime and the mud. Which is the more honorable material, — mysterious, complex life, or dirt? I should be ashamed of arguing this point, were it not for the prejudice. I am not half as anxious to find out that I did not come from an ape, as I am to know that I am not travelling toward one. Where we came from, touches not the matter of what we are. "Now are we the sons of God;" and, if evolution be true, well and grandly may we add, "It doth not yet appear what we shall be." People who get up in the world are sometimes ashamed of their parentage; but I think it much more important that we be careful that our children have no reason to be ashamed of their parentage. And since my line runs back millions of years, and ends in God, I see no good cause for being ashamed of the long and wondrous way by which it has come. Say it plainly, then: we have derived our present life from the animals. For the forces and laws of this development, I must refer you to the works devoted to their explanation. It is sufficient for me to say that known laws and forces are able to account for all the facts and results of evolution. As an illustrative hint of the power of nature over humanity, you may take it as an approximately correct statement, that if a race could start at the North Pole, and march southward, being some millions of years on the journey, stopping long enough for all the forces of its changing conditions to produce their effects upon the new-coming members, such a race would pass through all the changes, and exhibit all the peculiarities, of all the races that have been actually subjected to these various conditions. The natural force of development worked on the body until it reached its upright attitude and present comparative perfection. Then, when brain-power became the winning element in the struggle for life, the same force turned to brain. And now the moral is gradually gaining supremacy; and the time will come when this will be reckoned the mightiest, as it is now admitted to be the noblest, force of humanity. The moral power is coming to be the power that wins.

Thus, in accordance with these hints, evolution has no more difficulty in accounting for intellect and righteousness, than it has in explaining muscles and bones.

Let us now look at some suggestions and probabilities.

And, first, there is a marvellous sense of sympathy in our souls for the whole wide life of nature. Lowell talks of trees being his ancestors, and of his lying under the willows and listening to their whispers. Who has not felt the storm, or the stars, or the pines, speaking to his inner consciousness? By more than poetic figure we call the earth our mother; and we love, in sunny spring, to go back to our childhood, and lie upon her bosom. Every sensitive nature has a subtle sense of kinship with all the forms of natural life. Nature plays upon us, and creates our moods. We sing with the birds, and cry with the rain. The sunshine or gloom of the sky gets reflected in our faces. The life of spring starts our blood into tingling pulsation in unison with the waking activity of animals and the sap in the trees. Now, I believe all this wondrous feeling of kinship is best explained by saying we are akin.

And if you will notice the particular facts of animal life, you will be surprised that the gulf between them and us is so very narrow. What was once supposed to be an impassable ocean is only a tiny rill. The animals share with us almost every one of our habits and faculties. Animals think, reason, hope, fear, remember, laugh, and cry. They are faithful to their marriage-bonds, and devoted to their offspring; they are organized in communities and governments; they domesticate and use other animals, even milking them as we do cows; they have social grades, aristocracies, soldiers, artisans, and slaves; they make war, and bring home captives; they punish for crime, and execute the death-penalty; they lay out and build cities; they station sentinels, and have watchmen; they decorate and ornament their persons; they care for each other, and will fight in each other's defence, or in the common defence of home and country: in short, there is hardly a single faculty of humanity they do not share. The difference between these animal powers and customs and those of man is chiefly one of degree. The power of abstract thought and the development of the religious faculty and life are the great essential intellectual and spiritual distinctions. The animal world, then, is only a step beneath us; and that step evolution is perfectly able to take.

Coming to and dealing with the body alone, it is a significant fact that almost all forms of life have what are called rudimentary organs; that is, certain limbs or parts that were fully developed in one grade of life, but, not being needed in the higher grade, are outgrown, leaving behind them only a rudiment to show where and what they used to be; just as, in an old tree, you often see where a limb once was that has now died out. The fishes in Mammoth Cave have rudimentary eyes. Ostriches have rudimentary wings. These as simple illustrations. I can only stop to say that man also has several very striking rudimentary organs, which, if developed to-day, would make him as thoroughly animal in his appearance as is even his chimpanzee ancestor.

Another very remarkable thing I must only mention. Every child, before its birth, in the course of its development, passes through every phase of animal life from the lowest to the highest. Thus man recapitulates and takes up into himself all the life beneath him. Every infant born passes over again the whole pathway of the progress of life on the globe.

And then, after he is born, every child begins life an animal, and grows up through and out of barbarism into the civilization of culture and training. The child's playthings are copies of barbaric weapons, — club, and bow and arrow, and spear; and the games of children in the nursery and on the sidewalk, and the school playground, are mimic copyings of old religious rituals; and their meaningless rhymes and formulas are the remnants of Old-World stately ceremonials. So the child again, in his training, recapitulates and lives over once more the whole progress of civilization.

We are accustomed to say that such a man is foxy, another is lion-like, a third is wolfish, there are "bulls and bears" in the stock-market, others are swinish. These are looked upon as purely figures of speech; but evolution fills such phrases with a meaning they did not have before. Just as we may have the hair or eyes or gait of not only our father or mother, but of ancestors a hundred generations gone, so we may show still the good or evil traits, propensities, and passions, that characterized our animal ancestry a thousand ages ago. The thread of a common life runs through, and binds together in one, all forms of existence on the planet.

The battle of our moral life, on this theory, is rationally explained to be just what we know it to be, — a fight between the higher and the lower. The animal obeys his impulses; and, having no moral sense, of course there is no thought of wrong, nor any possibility of remorse. The child, at first, does the same, and has no more moral sense than the horse. But as this sense unfolds, the conflict begins. It is the contest between impulses, and duties toward our fellows and our higher self. "I wish" and "I ought" are in antagonism. It is just the fight that Paul so graphically pictures in the seventh chapter of Romans: "I delight in the law of God after the inner [higher] man; but I find another law in my members [body] warring against the law of my mind." And, as many of us have exclaimed, he cries out, "O, wretched man that I am! who shall deliver me?"

This theory, then, adequately explains the whole battle of the moral life. Man is struggling up, out of the animal, toward mind and spirit. When the animal gains the mastery, he is degraded, and falls back into a position worse than that of the animal, by as much as he is capable of something higher; and so when he misses it he is self-condemned, and condemned by mankind. However low or mean, you do not condemn a thing if it is all it is capable of; but, however high, you still condemn if it is content to fall below its highest possibility. Sin, then, is not a substance or entity that either god or devil put into a man, and that a priest can exorcise, prayer expel, or baptism wash away. It is only the supremacy of the lower over the higher in man; and guilt is the consciousness of this. Righteousness is the supremacy of the moral and spiritual, — a rational discernment of the laws of our higher life, and an obedience to them.

But if man was once animal, and has grown up into man, when and how did the soul come in? some of you may like to ask. Of any soul that is a distinct and separate entity, apart from the conscious mental and spiritual life; a soul that a man *has*, and that can be *saved*, apart from his mental and moral condition, according to the teaching of the popular revivalists; a soul that is in a man and yet not simply and wholly himself, — of such a soul I must confess that I know nothing whatever. And if any one is disposed to be troubled on this point in connection with evolution, perhaps it is well to remind them that they will find no relief in Genesis. Moses knows nothing of any such soul. The Hebrew word for the soul of Adam, and for the souls, or life, of the animals, is precisely the same. When it is written, "The Elohim breathed into his nostrils, and he became a living soul," it would be just as correct to say, "He became alive, or a living being or animal." There is no hint that his soul was any different from that of any other creature's soul. This does not touch the question of the nature of the soul, or of immortality: it only shows that there is no more light in Genesis than there is in evolution.

It was a favorite topic of discussion among the schoolmen of the middle ages, as to whether we derived our souls by birth from Adam, or whether they came direct from heaven at every separate birth; but I have never heard that it resulted in any thing more profitable than that other question, which they also expended their wits upon, as to whether our hair and finger-nails, that we now cut off, will or will not form a part of the resurrection body.

Evolution simply takes man as it finds him, traces his origin, studies his nature, and, looking along the line of his development, attempts to forecast his probable future. If it cannot say as broadly as Paul does, "Now are we [all men] the sons of God," it can say, "Now are all who live after the law of their higher nature the sons of God." Man, then, once animal, has climbed up into the possibility of sonship to the Highest; and in many cases he has made the possibility a fact. "It doth not yet appear what we shall be;" but along the line of knowledge, obedience, and struggle, there stretches before humanity an ascending pathway up to God, bright and grand as the sloping beams of light that bridge the deeps of space from the horizon's edge up to the unbearable glory of the rising sun. Here lies the way of religious progress, philanthropic labor, and all reform. It is to be trodden by all those who subdue the animal, and climb up into the mind and the soul. And the true help for our fellows is to be offered in assisting them up, step by step, along this same stairway of attainment. No man can be suddenly "converted" into it, any more than ignorance can be suddenly converted into knowledge. One may be suddenly waked up to the fact of his ignorance, and may suddenly determine to go to school; but the way of learning is long, and so is the way of all progress.

In the light of human nature, as thus revealed, may be seen the futility of some of the present prominent notions about reform, as in matters of temperance, social vice, and the repression of crime. You cannot legislate character. The most that laws can do is to help on suitable conditions for the development of character.

I have purposely passed by the question of a future life for man. It only remains for me to glance at the possibilities of his earthly outlook, as hinted by evolution. Man was first an animal, without shelter or fire or clothing or weapons or domestic utensils of any kind. We can trace him to the time when he fought the bear for possession of his cave. His first weapon was a club; then he discovered fire, and so was able to mould metals, and manufacture utensils and weapons. All the forces of nature were hostile to him, because he was ignorant of their laws, and so was constantly transgressing and suffering. But gradually he learned to obey them, and so became their master. When he obeyed the laws of the wind, he made it sail his boats; when he knew the laws of water, he made it turn his mills; when he learned the use of fire, he cleared the forests and discovered manufactures: to-day he has made conquest, by his knowledge, of immense tracts of the globe. Lightning, light, heat, magnetism, chemical forces, are all become his servants. He is just winning his crown and grasping his sceptre. But though we call ourselves civilized, we can see enough of the yet unattained possibilities of man and the earth to make us feel that we are as yet only in the morning twilight: the full day is before us; conquests await us in earth and air and sea. Government shall be perfected;

crime shall be outgrown; most of the diseases from which we now suffer shall be abolished. Accumulated wealth, and knowledge of the yet undiscovered resources of the earth, shall solve the problems of hunger and cold and want, and deliver man from the crushing burdens of the mere struggle for subsistence. Then man will be free to go up and live in the affections, the mind, and the spirit. It is perfectly within the scope of the forces and laws now at work about us, to develop on earth a paradise as much fairer than Eden as the noble plans and works of manhood are higher and better than the dreams of the nursery. Plato's republic and More's Utopia are only hints of what the future will realize. The whole earth can be made a garden, and human life upon it so regulated in accord with natural laws that all government and society and individual life shall run as noiselessly and harmoniously as the stars move in the sky. Evil and pain and disease will be outgrown.

How long will it take? Thousands of years. But God is in no hurry, having eternity to work in. He has been millenniums already, in developing us to what we are; and yet we are only on the threshold of what is not only possible but probable. The future stretches out a road so long, that those who stand on the heights of the ages to come may look back and think of human history from the first until now, as we think of the first mile of a journey across the continent; as we regard the time from Adam to the flood, in the old story.

"Beloved, now are we the sons of God:" and indeed

"it doth not yet appear what we shall be." But the scroll of earth-life that God is unrolling has in it wonders and surprises of good and beauty and glory that we cannot now even imagine; but we may safely say that the blossom will be worthy of the root and the trunk.

V.

THE DEVIL; OR, THE NATURE OF EVIL.

To the devil has been popularly attributed the authorship of all human ills. I propose, therefore, to trace in rapid outline the more important phases of the development of the satanic idea, that we may understand how the belief has risen and grown. Then I shall pass to what we really know of the origin and nature of evil. A knowledge of what it is will suggest practical measures for controlling or escaping it.

Let us begin by trying to understand the condition and state of mind of the first men on the earth. You must imagine yourselves divested of every thing that we include under the term "civilization." Not only must you mentally blot out the cities, the telegraphs, the railways, the roads, the ships, but you must also think of them as without houses, without even the simplest domestic implements and utensils, without any knowledge of iron or copper or other metals of which they could make weapons for their defence. They had no clothing, except as they could strip a tree of its bark, or rob an animal of its skin.

They possessed no knowledge of fire. And then you must not only strip their bodies of all the surroundings of comfort: you must also strip their minds of knowledge, and picture them in the midst of a world of which they knew almost nothing. Hunger gnawed at their vitals, the cold chilled and destroyed them, the thunder scared, and the lightning smote them; the earthquake, the storm, and the darkness all seemed bent on their destruction. Invisible diseases attacked them, and mysterious death snatched away one after another. They knew nothing of natural laws, or of how a generally beneficent force might, because of their ignorance or disobedience, work them incidental injury. So they, using the best powers of reason they possessed, argued that all these things were their enemies. And as they had no idea of power apart from that possessed by persons like themselves, they attributed personality and individual life to all the forces about them. Winter and night and storm and lightning and disease, to their thought, became superhuman beings, or gods. And since the results of their action were evil, so far as they could see, they concluded they must be malicious gods, inclined not to help, but to hurt them. The sun and the light and the blue sky were kindly and good gods. For not knowing, as we know to-day, that all these apparently evil forces were the result of the sun, they gave him credit only for the pleasant effects of light and warmth which they could see him produce. But the good gods of light and comfort were afar off in the sky; while the bad gods of cold and darkness and hunger and death were right about them. Thus the first religion was one almost exclusively of fear. They worshipped these gods, not because they loved them, but in the hope that, by gifts and prayers and honors, they might gain their favor, or induce them to be less cruel. Here, then, is the germ-idea of the devil. He was born of the logic that argued that suffering and death must be the work of a wicked being. It was not one devil, or bad god, but a thousand; for every thing that seemed to possess independent power was personified.

In the Hindoo religion, though almost every thing was deified, so that there were millions of gods, they yet took one step ahead of the philosophic thought of primitive man. They generalized the forces of the universe under three grand gods, — Brahma the creator, Vishnu the preserver, and Siva the destroyer. The latter, who may be called the Hindoo devil, was figured with a rope for strangling evil-doers, a necklace of skulls, and earrings of serpents. All evils were of his doing. And, as very significant of the fact that popular worship concerns itself much more about escaping evil than it does in getting good, Siva is the most worshipped of all the Hindoo gods. The idea seems to be to buy escape from his evil power.

Zoroaster, the religious teacher of Persia, took one step more in advance. His religion was based in a shadowy kind of monotheism; but practically the universe was divided between two gods, a bad and a good one, who were supposed to be in everlasting antagonism. Ahura-Mazda was the god of light and good, while

Ahriman was the god of darkness and evil. They were engaged in a world-wide conflict very much as, during the middle ages, Christ and the devil were supposed to be. But, unlike the Christian teaching, it was believed that sometime Ahura-Mazda would triumph even to the utter destruction of all evil, and the redemption of Ahriman himself.

It is worth your while to notice, in passing, how the gods of an old religion become the devils in a new. When a new and better god got the supremacy, it was not supposed that the other gods were dead: they were simply conquered and kept in subjection. In this condition they kept up a chronic rebellion, and did all they could to injure their conqueror and his kingdom. Thus the devils of Zoroaster were the gods of the older Hindoo religion. This is indicated in the very word "devil." *Deva*, from which the word "devil" is derived, once meant the good gods. The same use remains to-day; for "divine" is only the old word *deva* in its modern dress. So "devil" and "divine" are two words coming from the same root. The same is true of the Latin *deus* and our English "deity." In like manner the old Pagan gods became the devils of early and mediæval Christianity; except when some infallible pope put one of them into the calendar by mistake, and made him a saint.

In the classic religions of Greece and Rome, there was no devil; for the gods were bad enough to get along without one. The deities of Olympus were simply human, except as to power and life; and they helped or hurt humanity, as all tyrants do, according to freak or favoritism, or as moved by gifts or honors.

We must now turn to the Old and New Testaments, and trace the genesis of the devil here. There is no hint, for ages after, that the serpent in Eden was the devil, or that he had any thing to do with him. Jehovah, at first, as being the one God, the Creator of all things, is regarded as the author of both good and evil, dark and light. In Isaiah he is represented as boldly assuming the whole responsibility to himself: "I form the light, and create darkness; I make peace, and create evil. I, Jehovah, do all these things." (Isa. xlv. 7.) This, of course, could not be true if evil originated with the devil. In other places, the agency of the devil appears springing up, and dividing the evil work of the world with Jehovah. When David committed the supposed crime of taking a census of his people, it is said in one place that Jehovah tempted him to do it, and in another place that Satan did it. The belief here is divided, and begins to waver. The first time that Satan appears clearly outlined is in the book of Job. But he is not at all the Satan of theology, outcast from heaven, the king of hell, and the ravager of the earth. He is a member of the heavenly court, a sort of prosecuting attorney for Jehovah, who goes forth only on the divine permission, and to execute the divine will. But after the Jewish captivity, and the national contact with Persian life and thought, the Zoroastrian Ahriman comes bodily into Jewish theology, and the devil is full born.

To account for the existence of evil at the dawn of creation, the devil must be made older than the world; and hence arose the legend of a rebellion in heaven, and the casting-out of Satan and his followers, a third part of all the angels. To avenge himself for his celestial defeat, he turns his malice against the new creation and the new being, man, who was to take the place in heaven from which he had been cast out. Then arose the idea that the old serpent was only Satan in disguise; and that, through the tempting and fall of man, the devil "brought death into the world, and all our woe." The devil was thus supposed to have become the conqueror, lord, and rightful owner of the world. This was his kingdom. So ingrained did this thought become that even in the Church Prayer-Book to-day the organized Church is God's kingdom, and the world is Satan's; and becoming a Christian is "renouncing" loyalty to "the world, the flesh, and the devil," and taking Christ as king. So perfectly was it believed that Satan had come into ownership of earth and man, that the early doctrine of the work of Christ was shaped by it. Jesus made a bargain with Satan, by which he surrendered himself into his hands as a ransom for the deliverance of humanity; and as Christ had been his rival in heaven, he was willing to accept the purchase. But Christ, being divine, was able to outwit the devil, cheat him out of possession of himself, and save humanity into the bargain. Such were the early Christian ideas of God and Jesus and Satan. You can judge how worthy they are of our respect.

Throughout the middle ages, the devil was supposed to be everywhere. His armies camped over all the earth. Calamities, sickness, deformity, earthquake, storm, down even to the most trivial perplexities, were attributed to his agency. Every man was surrounded by devils, as a swimmer was surrounded by the sea. Evil thoughts were his whispers, and evil deeds were done at his instigation. In short, whatever was evil was the work of the devil. Sin was a new element or quality that came into human nature, through his agency, at the fall. Thus it became a part of the life of every new child born into the world. By this sin, which linked man with Satan, he was under the perpetual wrath of God, and condemned forever. The Church ordinances, rituals, days, — these were the machinery of Christ for the expulsion of sin, or the deliverance of man from its power.

The devil has been the mainspring of theology, and hell the corner-stone of the universe. So important a personage has he become, that one minister asserts, as against Mr. Conway's lecture, that "if there is no devil there is no truth in the Bible." As if the Sermon on the Mount must look to the devil for support! And another declares that "the devil constitutes an important element in the means of grace established for the salvation of sinners;" though, since he gets the largest part of them, one would suppose that it was their damnation, and not their salvation, of which he was an "important element." I suppose this minister means that he needs the devil as a scarecrow, to frighten people into heaven; though, by the Church's own confession, he has proved a failure in this direction.

Now, it is perfectly natural that the ignorance of primitive man should invent the devil as a part of his mythology. It was the simplest explanation of the facts of evil, as they then appeared. But it is now being seen by all earnest and independent thinkers, that the theory of the devil must take its place with alchemy, the Ptolemaic theory of the universe, and other beliefs that the knowledge of the world has outgrown. The devil only complicates, instead of simplifying or explaining, the origin and nature of evil. If there is any devil, he either made himself, or God made him. If he made himself, he is an independent deity; if God made him, or permitted him, then God is the author of evil: so the devil is no relief.

Let us, then, turn away from these frightful dreams of the past, and look at the facts of the world, so far as we know them. I think we shall find, in the light of evolution, a pathway out of our difficulties. What does our real knowledge of the facts of the world and the progress of life tell us about the nature of evil? It denies not only the existence of any devil, but any necessity for him. It denies the real existence of evil; that is, it denies that evil is a real entity, a substance, either in the world about us, or, as sin, in man. Evil is simply a temporary and passing condition. To put the whole thing in one word, all evil is nothing more nor less than maladjustment. The devil, and sin, and sorrow, and calamity, and sickness, and tears, and death, all resolve themselves into this one word. I am aware that as thus stated the sentence may convey to you little meaning; but I wished to put the whole thing into one word, as an acorn contains an oak-tree. And now I will go on to explain and illustrate. If you can find any form of evil that cannot be wrapped up in the word "maladjustment," then you will find what all my thinking has failed to discover.

At the outset of this explanation, notice the meaning of life. Life means simply this: A real being, man or woman, in the midst of, surrounded by, and related to, the real facts of the universe. When he is rightly related or adjusted to these facts, then the man finds safety, health, happiness, and peace. When wrongly related, or maladjusted, he finds disturbance, pain, calamity, sickness, and death. This one principle, I believe, will explain all evil, from the physical up through intellectual and moral to the spiritual. Let us now start out with it on our search, and see.

I. Physical Evil.

How came it about that the first men of the world interpreted all the great forces of the world as being mainly evil? It was simply because of their ignorance and helplessness. Not understanding the laws of nature, they were continually breaking them, and so suffering the penalty of broken law. The lightning, the storm, the earthquake, the winds, the poisonous vapors of swamps, all these were continually working them harm because they had not enough knowledge of them to be an adequate defence. They were constantly getting into wrong relations to these things. The great forces of nature move on as resistlessly as a train of cars under full headway. If you get aboard the cars, the mighty power of the steam will speed you smoothly on your journey: if you get across the track, the same mighty power will crush and mangle and kill. The whole thing depends on your relation to the force of the steam. Right adjustment makes it your servant: maladjustment makes it your ruin. In the light of this principle, let us consider more directly some of the prominent causes of physical evil.

It has been the popular belief of the Church and of all Christendom, for the last two thousand years, that earthquakes and tempests and floods and volcanic eruptions and pestilences and diseases and death were all either the spiteful work of the devil, or else the vengeful judgments of God for human sin. Sometimes the two ideas were combined; that is, it was thought that God, being angry, permitted the devil to bring these evils upon men. This belief was carried so far that it was even held to be an impious resistance to God's will, when medical men endeavored to counteract the natural sweep of a pestilence, or to alleviate human suffering. Within a hundred years, the world has witnessed the advocates of the popular religion roused into a storm of opposition against vaccination as a preventive of small-pox. Smallpox was a judgment of God; and it was wicked to stand in its way. And, so late as forty years ago, the use of chloroform was opposed because it was thought to be impiety to fight against the pain that God had sent as a punishment for sin. Now, let us look at it, and see.

Earthquakes and volcanic eruptions are a part of the natural and necessary process in the development of the globe. The wrinkling and tremors of the earth's crust, in the process of the earth's cooling from its original molten condition, are as necessary and natural as the wrinkling and cracking of the surface of a pan of lard when it is cooling in the farmer's kitchen. It is precisely the same thing, only on a larger scale. And volcanoes are the natural result of the same working. Now, when, and in what sense, are they calamities? When men, through ignorance of these natural laws, build their cities or homes in the earthquake's track, or on a volcano's side, then there come such calamities as that of the overthrow of Lisbon, or the burial of Pompeii. These great convulsions are, on the whole, beneficent, for the good of the earth; the incidental evils are the result of a maladjustment of human relations to them. So soon as men know the laws, and adjust themselves to their working, the power to injure is taken away. Take the case of floods, or the breaking of a great reservoir, like that in Western Massachusetts a couple of years ago. When men know enough to build out of the way of floods, and when the public guardians know enough and care enough to prevent the cupidity of contractors from slighting their work, then these "judgments of God" will be found easily preventable. It is a little curious that God has ceased sending his lightning as a punishment for sin just as fast and as far as Franklin's invention of lightning-rods has been properly applied. Right adjustment to electricity has disarmed it. When a ship is wrecked at sea, we know that, in every case, it is for lack of proper adjustment between the vessel, and the powers of wind and wave. A proper knowledge of, and a compliance with, the laws of navigation, would make wrecks almost an unheard-of thing. Pestilences, like the cholera, the plague, the yellow-fever, are perfectly within the control of human intelligence. These punishments come not on sinners, in the ordinary sense of the word, but confine themselves exclusively to those cities and lands that disregard sanitary laws. And every intelligent physician knows that almost all diseases are preventable, and that none ever come except as the natural and necessary result of the disregard of physical laws. In other words, all disease is maladjustment between man and nature. He who builds his home beside a malarious swamp may be ever so pious, but he will shake with chills and fever just the same. The wicked man who knows enough to keep out of such places will escape. That is, transgression of physical law brings physical evil; and what you do or do not do in spiritual matters does not touch this fact. Fire is one of the grandest blessings of the race; but if Mrs. O'Leary does not know any better than to put a kerosene-lamp where her cow will kick it over, the laws of fire are not going to stop: Chicago will crumble into ashes. A man's house tumbles about his ears; not because gravitation is not a good thing, but because he did not adjust his timbers and his bricks to the laws of gravitation. And so you may take any illustration you please, and you will find the same thing true : all physical evil is maladjustment between man and his conditions. Increasing knowledge is disarming these enemies of humanity more and more. And not only are they being disarmed: knowledge of and obedience to these great laws is turning them into gigantic servants of the welfare of man. The sea, which the ancients regarded as a waste of useless waters, — the separator of nations, the destroyer of those who braved its perils, and which even the Bible speaks of as an evil to be done away in "the new heavens and new earth," — " there shall be no more sea," — this is now known to be the life of the world; and knowledge of its laws is turning it into highways of human progress. The lightning has turned mail-carrier; and all the great forces are being harnessed to the car of human advancement. There is not a single physical evil on the globe, that future knowledge may not turn to man's advantage and uplifting.

Death, the great evil, as men have thought, and which was supposed to be the result of Adam's sin, is now known to be as natural and necessary as life. It existed ages before Adam is supposed to have been created. The evils of it are only incidents and accompaniments that human advancement will leave behind. The time may and will come when men will die as unconsciously and easily as now they are born. In fact, it will only be another and a higher birth.

The devil of the physical world, then, is only ignorant maladjustment. This is the great dragon; and knowledge is the St. George that will overthrow and slay him.

II. Intellectual Evil.

I pass now to say that the same is true of intellectual evil. This calls for no prolonged discussion. Its simple statement will be its sufficient proof. The only evil of the intellect is error, a lack of true knowledge of the real facts of the universe. The only and the sufficient cure for this is knowledge. The straight pathway to this knowledge is the free and unhindered study that cares only to find the truth. This is why I claim and defend the right of free inquiry. Nothing is safe but truth; and truth is always safe. And I can conceive no interest that any man can have to do any thing else but find the truth, and obey it; for truth is God. The man who builds on it builds on the rock. The rains may descend, the floods may come, the wind may blow and beat upon his house: but it will stand; for it is founded on a rock. Any thing but truth is sand; and whoever builds on it will live only to see the ruin of his fall.

III. Moral Evil.

As our third point, consider the same principle of maladjustment in relation to moral evil. The unfolding history of humanity reveals nothing more plainly than that there are great and universal laws of righteousness running through all the world. These laws are nothing more nor less than the all-inclusive principles of equity or just relationships in which men stand to each other. Within these lines or laws of right is moral goodness. Breaking over these lines or laws is moral injury and destruction. Here, as everywhere, "sin is the transgression of the law."

Now, I wish you to carefully separate in thought this realm of morals, of righteousness, from the intellectual and physical realms. Breaking physical law brings physical evil; breaking intellectual laws brings intellectual evil, or error; breaking the laws of righteousness brings moral evil, or the injury of the moral nature. These three realms may run into each other. As, for example, when a man breaks the moral law of his relationship to another by some sensual crime, he at the same time may break a physical law of his body; but the physical evil results, not from breaking the spiritual law, but from breaking the physical. A man may be a thief, or cheat in his business, and it have no effect on his body. He will be in health so long as he obeys the laws of his body, however much he may break the higher laws. The farmer may swear and cheat; but this will not affect his potato or wheat crops, if he is a good farmer. While another farmer may be ever so noble and true in his relations to his fellow-men; still he will go hungry if he does not manage his farming wisely. The higher moral evils, then, result from disregard of the higher moral laws of equity in which we stand to God and our fellow-men. Thus society may be full of people who are successful in their outward circumstances, because they know and obey the laws of health and of business. They may be successful in the pursuit of the scientific and philosophic truths of the world, because they obey the laws of this search; and at the same time they may be so out of harmony, of right relationships, with the spiritual laws of righteousness, that in this department of their being all is only wreck and ruin. A man will be a good painter, if he obeys the laws of his art; but this will not teach him music. He may be a fine mathematician; but this will not teach him sculpture. So, if one is to be morally and spiritually developed and complete, he must recognize and obey the higher and finer laws of his higher relationships. When evolution has carried man high enough, this spiritual realm of relations will be as real to him as is now the physical; and the pang of pain following a breaking of one of these finer laws of thinking and feeling will be as keen and real as is now the burning that follows the putting of a finger into the fire.

Moral evil, then, is only moral maladjustment, — a man getting out of right moral relationships to God or his fellow-men; and the cure for evil here is parallel to the cure for all other evils. Man universally desires his own welfare, that which is for his good; and when he gets wise enough to know that obeying the divine laws must always be for his good, then the whole force of self-interest even will be turned toward doing right. Man knows it now concerning physical evils. He does not voluntarily seek them. They come through ignorance or weakness or carelessness; or sometimes he braves them for the sake of what he thinks a larger good. He is beginning to learn it concerning intellectual evil, or error.

He will one day become wise enough to know that it is always safe and for his interest to find and follow the truth. But he has as yet learned very little of the same great principle as applied to moral relations. I think no man ever does wrong at first, unless he has come to think it good for him, for his interest at the time. This is always a blunder; but he does not know it yet. For it is for no man's real interest to do wrong. But the selfish hunger is strong, and his moral sense is weak and blind. The sense of duty is only partially developed. Habit grows until it becomes disease; and so the current sweeps him away. But a proper knowledge of moral laws, at the start, would have kept him out of the current.

Knowledge, then, is the devil-killer, and the exterminator of evil. It is often objected to this teaching (it is, indeed, the stock-argument of religious newspapers), that many educated men are criminals, such as Ruloff and Green and Prof. Webster, the murderers; and therefore it is said that education cannot save men from sin. But a fatal fallacy underlies the objection. It is not claimed that physical education will save from any thing but physical evils. Knowledge of scientific and philosophic principles will only save from errors in their peculiar realm. It is moral education, a true and adequate knowledge of moral laws and human well-being as related to these, that saves from moral evil. Ruloff and Webster and Green were moral fools. They blundered concerning the great moral relations in which they stood, and mistook their own real welfare, as well as the rights of others. When men are morally wise enough, they will know it is always best to do right.

The perfect humanity will come, then, when there is a complete knowledge of human relationships, and a complete obedience to the physical, mental, and moral laws of God.

God, then, is not the author of evil. It is only human maladjustment to laws that, when known and kept, are the true servants and the mighty helpers of humanity. The devil is a dream of the night and darkness of the past. Let him be relegated to the museum of theological curiosities, mummies, and skeletons, that the coming ages will study to find out the world's thoughts that have passed away. But meantime remember that unkept laws are still capable of doing the works of destruction and sorrow and ruin that have been credited to the devil in all time. If physical, intellectual, and moral calamities come, it will be small comfort to you to know that your own ignorance or carelessness is the devil that brings them. But in this destruction of the devil, and of a wrong-doing God, we get the grand inspiration and the hope that helps us face the future with good heart and tireless endeavor. The way out and up being seen, and a glimpse of light being visible as a herald of the dawn, we "thank God, and take courage." Let our hope sing itself in the lines of Tennyson: —

> "O, yet we trust that somehow good
> Will be the final goal of ill,
> To pangs of nature, sins of will,
> Defects of doubt, and taints of blood:

"That nothing walks with aimless feet;
  That not one life shall be destroyed,
  Or cast as rubbish to the void,
When God hath made the pile complete;

"That not a worm is cloven in vain;
  That not a moth with vain desire
  Is shrivelled in a fruitless fire,
Or but subserves another's gain.

. . . .

"I can but trust that good shall fall
  At last — far off — at last, to all,
  And every winter change to spring."

VI.

THE EVOLUTION OF CONSCIENCE.

You all remember the story of the boy Theodore Parker; how, when out in the field one day, he lifted a stick to strike a turtle, but stood startled, and holding the stick suspended in the air, because he seemed to hear a voice distinctly saying, "It is wrong." On returning home, he asked his mother what it meant; and she replied, "It was the voice of God in your soul." And then she went on to explain that, if he always listened to and followed it, it would guide him in the way of right; but, if not, it would fade out, and leave him in darkness and error. Now, this might all be true with the conscience of the boy of Theodore Parker's mother; for she was exceptionally intelligent and religious. But a little observation would have discovered some remarkable exceptions in the case of other boys in the neighborhood. Many of them would have struck every tortoise they came across, or have put coals on their backs to see them run, and never have heard any voice of God say any thing about it. What Theodore Parker thought a sin, they would have declared to be only jolly fun. What I wish to hint is simply this: that this conscience, whatever it be, is not uniform in its utterances, and does not declare any general law of action. It tells one person that a thing is wrong; it tells another that it is right; it tells a third nothing about it.

These words of Mrs. Parker undoubtedly express what has been the popular doctrine of Christendom. Conscience has been supposed to be a direct intuition of right, or the voice of God in the soul telling a person what was right and what was wrong. But let us look a little over the world, and see if we can still hold this belief. The first thing that we observe is that the utterances of conscience in different parts of the world not only are not all alike, but that they are even contradictory. Hardly a question concerning God, the state, society, or the individual, that you cannot find conscience engaged on both sides of. In one nation, conscience commands the worship of Jehovah; in another, with equal imperativeness, it commands the worship of Jupiter, or Brahma, or Moloch, or Venus, or a bull, or a stone, or some sacred serpent. In the breasts of our forefathers, it commanded resistance to tyrants, and a contest for human liberty, even at the cost of property and life. In Turkey, conscience makes it a sacred duty to heaven to be abjectly submissive to a most degrading despotism. Conscience, in the better parts of Europe and America, commands a state of society in which one husband is faithful to one wife, and in which both live for the welfare of the children.

In Utah, conscience binds two or forty women to one man. In other parts of the world, this same conscience sanctions the possession by one woman of several husbands. The conscience of Plato led him to recommend an ideal community, where husbands and wives should be held in common. The American conscience commands the most tender care of children, and the watching over and securing comfort to the declining years of parents. In India, conscience bids the loving mother cast her child to the crocodile; and in the South Seas it commands children to strangle their parents, or bury them alive. It is held a religious duty to do this before they get old and decrepit, that thus they may enter their eternal life young and strong. Conscience in Sparta commanded people to steal. Conscience in America builds prisons for thieves. Conscience used to forbid taking interest on loans. Money-lenders now worship God, build churches, and sustain missions on the interest of their capital. Conscience once carried on the slave-trade, and established the institution in America. Conscience at last rose against it, declared it "the sum of all villanies," and poured out blood and treasure to overthrow it. Conscience hung John Brown; and it made John Brown willing to be hung. Conscience crucified Jesus; and conscience deified him because he was willing to die for his ideal of right. Conscience built the Inquisition; and conscience made men strong to bear its tortures. Thus it appears that conscience gives contradictory statements, and is not, by any means, the same all over the world.

There is hardly a namable crime that conscience has not somewhere consecrated as a duty, and there is hardly a namable duty, that conscience has not somewhere condemned as a crime; and not only is this true of the condition of the world to-day, but there has been a continual change and progression in the doctrine of right and wrong, along the line of the pathway of human history. The conscience of the highest civilization is as much above the lowest, as the highest types and forms of life on the globe are above the fishes and the reptiles. But the "voice of God," if it utter the constant and eternal truth, cannot progress. The conscience of each man also changes and develops in accordance with his own conditions, education, and development from childhood to age. He conscientiously drops some things he once regarded as duties, and he feels bound by new obligations that before he did not recognize.

In the midst of all these changes and fluctuations, there has been one and only one thing that has remained constant and unmoved, like a rock in the midst of the sea. This one thing is the fact that all men everywhere have recognized and acknowledged the distinction between right and wrong, and have confessed the meaning of *ought*, — ought to do right, ought not to do wrong. Disagreeing as to what was right, and what was wrong, they have yet admitted that they ought to do the one, and avoid the other. This basal rock we can rest on, and make it the foundation on which to build a true doctrine. But we are not quite ready for that yet.

I wish you to notice here that this universal conflict and contradiction — a very Babel of consciences — utterly disproves the teaching that the human conscience is, in any exceptional sense, the "voice of God in the soul." It is a purely human faculty, like the faculty for art or music; and it gets its authority, as they do, by being true, and just in so far as it is true. Now, orthodoxy regards these other powers as purely natural; but it makes conscience an exception, regarding it as a sort of permanent inspiration, "inner light," or dwelling of God in the soul. But if God in the soul speaks like a hanger-on at the White House, or a paid partisan, according to circumstances, it is hard to see what advantage there is in calling it divine. The important question is, Does it speak the truth? Orthodoxy explains the fallibility of conscience by the fall of man, saying that this once divinely infallible faculty is now only a part of the general wreck of humanity; but this is only another way of confessing that the accuracy of conscience depends on the individual condition, training, and character of every man. And on this supposition, the consciences of all "converted" people — those recovered from the fall — ought to agree. But history teaches that the "saints" have hated and fought each other, for conscience' sake, quite as cordially as though they had been the chief of sinners. Evolution regards every faculty as an unfolding of the divine life of things, but conscience no more truly so than any other.

But from this confusion on the part of the utterances of conscience, many perplexed thinkers have hastily jumped to the conclusion that there is no real and ultimate standard of right, that it is a mere matter of convention. Were this so, every man and woman would be a law to themselves, and would be justified in doing " what was right in their own eyes;" and when a conflict arose between two individuals, or between the individual and the state or society, it could never rise to the dignity of a conflict of principle. There would be no principle; it would be on the level of a bear-fight. It would consecrate and make universal the doctrine that "might makes right." Any man, then, who should be willing to die for what he called right, would be simply a fool; and the great calendar of the world's saints and heroes and martyrs would be reduced to a list of candidates for a lunatic-asylum.

Conscience is not infallible: therefore, whether you call it "voice of God," or natural faculty, it practically comes to the overthrow of the old ideas concerning it. But the doctrine that right and wrong are conventional matters is still less tenable. Which way, then, shall we turn for the truth? The doctrine of evolution, as applied to the conscience, will be found able to explain all the facts, and solve all the difficulties. Let us, then, trace its root, and mark the line of its development.

The root-meaning of the word "conscience" hints its true significance. Conscience and consciousness are near relations. They come from the same stock, and differ chiefly in application and use. Consciousness is our own knowledge of ourselves and of the relations between our own faculties and powers. Conscience is our recognition of the relations, as right or wrong, in which we stand to those about us, — God and our fellows. *Con-scio* is to know with, in relations.

Now, let us see if we can find the basal principle of morals, the very root of the distinction between right and wrong. It is to be discovered in the fact of society, that man is not alone, but exists in certain necessary relations to others. If you will conceive a man utterly alone in the universe, and not only so, but stripped of all those faculties and powers that fit him for personal relations with others, you will see that such a man would be incapable of any moral action. He would not be moral, he would not be immoral: he would simply be unmoral. He would be incapable of doing, thinking, or suffering wrong: both rights and duties would be abolished. The first time that two persons looked each other in the face, and became conscious, or had a conscience, of the independent existence and rights of each other, then the first feeble moral sense became an element of human life. The lowest substratum out of which this sense was born was just this recognition of personal relationship or society. The germ of it may be found in the forms of life below the human, — in a bird's nest, a lion's den, or a herd of horses. Some highly developed dogs have been known to recognize the rights of their fellows, and to show a sense of shame for a mean action. But it was when primeval man developed the first rudimentary society, that the human conscience was born. It was the first feeble effort of the imagination to "put yourself in his place," and thus conceive that others were only other selves, having similar desires, and capable of feeling similar pains.

Conscience, then, began in, and was commensurate with, the first society. What was the first society? Naturally and of necessity it was the family. So at the first there was a family conscience. The father, the mother, and the children recognized certain relationships in which they stood to each other, and certain rights and duties as springing out of these relationships. But while having a conscience toward each other, they had no thought of any rights or duties as pertaining to a family of strangers. They could rob or murder or eat up another family with no compunctions of the moral sense or qualms of conscience whatever.

As the family developed into the tribe, there grew up, along with the widened relationships, a tribal conscience. All were bound together in the clan, by mutual rights and duties. And this tribal conscience has usually been so intense and strong that a member of the tribe could not possibly live in disregard of it. And yet, beyond the limits of the tribe, they have sometimes had no conscience at all. Take, for example, a clan of Scottish Highlanders. Robbery, ravishment, and murder were supposed good enough for the members of a hostile clan, while all these things were sternly repressed or severely punished within their own borders. Or look at the tribes of American Indians. A Delaware could be chivalrous and high-toned as a modern gentleman toward the members of his own tribe; and, at the same time, to lie to, to cheat, to ambush, torture, or scalp, an Iroquois, was a virtue.

Then, as the tribes grew large, organized, civilized, and settled in cities, there came the development of the city conscience. The palmiest days of Greece can illustrate this. The Athenian conscience was practically limited to Athens. The citizen of Athens, proud of his city and jealous of her rights, had no conscience toward Sparta.

But now that governments are not limited to city walls, when many cities and towns are leagued together under one national flag, we have developed a national conscience. The progress of civilization has hardly advanced beyond this as yet. Americans feel a wrong done to an American citizen anywhere on the globe; and they feel justified in urging the government to demand redress, even at the price of war. But they do not lie awake nights, nor get righteously angry, over wrongs done to a German or an Italian citizen. Yet in one case the wrong is as great and real as in the other. We feel an insult to the piece of bunting we call our flag, and are ready to fight for the national honor. But instead of having the same conscience of wrong when England is insulted, we can most of us remember the time when thousands of Americans would chuckle over it, and think it was good enough for her. We sympathize at once with the faintest scream of our eagle; but the sorrow changes to a grin of satisfaction when we hear the lion roar.

In regard to other and broader questions, however, we

bury the distinctions of nationality in the consciousness of the race.

In some directions there is what may be called a Caucasian, or white man's, conscience. It is only a little while ago, that it had passed into a proverb that "negroes had no rights which a white man was bound to respect." It was this spirit that Nasby caricatured when he proposed the sarcastic improvement of the New Testament words so as to make them read, "Suffer the little (white) children to come unto me." And to-day, in California, Christian men look on, and see the Chinese subjected to such treatment as is a disgrace to humanity. Such abuse of white men would kindle a revolution. Yet the San Francisco conscience is very comfortable over the matter.

The day will come when the present rudimentary notions of the rights and wrongs of man will be developed into all the breadth and grandeur of a comprehensive human conscience. All the old partial forms of conscience still remain, and have their representatives in other directions. In many the family conscience is still dominant; and the sense of right and wrong in their hearts grows weak as they leave the home. Then there is a church conscience. Even Paul could write, "Do good unto all men, but specially to them who are of the household of faith." And there is a club or corporate conscience, illustrated by such orders as the Masons. They feel obliged to help each other, not so much because they are men as because they are Masons. The church and the Masons are good, only they are partial.

So there are professional and business consciences. Ministers, lawyers, bankers, merchants, mechanics, each too often shut up their sympathies within their own limits, making one set of rules for themselves, and another for them "that are without." A true conscience will be wider than this. It will issue in what has been finely called "the enthusiasm of humanity." It will make binding on every soul the saying of Jesus, "Thou shalt love thy neighbor as thyself;" and the word "neighbor" shall be comprehensive enough to include every human being.

And one step more must conscience take. Paul could raise the question, and give a negative answer to his inquiry, as to whether God cared for oxen: "Doth God take care for oxen?" But that is a narrow and only partially developed sense of right and wrong, that leaves out of its account any living thing. There is a sentimental regard for animals, as well as for men, that is sometimes ready to oppose the justifiable sacrifice of a lesser and personal to a greater and general good. A few people care for pets. But the "Society for the Prevention of Cruelty to Animals" is a living witness for the fact that the human conscience is lamentably weak on the side of our dealings with the brute creation. English hunting scenes, Spanish bull-fights, as well as the practice of American sportsmen, witness the same thing. An old divine could speak of it as a proof of God's goodness to men, that he made the animals for their delectation in such delightful sports as bull and bear fights; and Macaulay said the Puritans in England opposed bear-baiting not so much because it hurt the bears, as because it gave pleasure to men.

Where there is any good reason for infringing the natural rights of the animal world, of course those rights are extinguished in a higher need. But I wish I could burn into every human soul the righteous indignation and contempt which the sensitive Cowper felt for any one who was mean enough to "needlessly set foot upon a worm." The rights of a worm are as sacred, in his degree, as yours are; and a true conscience will recognize them.

Now, you will notice that in this upward development and broadening of conscience it has been, all the way and everywhere, governed by the same one principle that we found to be its germ. It has everywhere sprung out of relationships, or supposed relationships; and it has been commensurate with a knowledge and sense of these relationships. Many real and vital relationships have been unperceived; and so in body, mind, and soul, men have suffered, lost, and died on account of this ignorance; but they have had no conscience about it. On the other hand, they have imagined a thousand relationships that had no real existence. But since supposed facts have all the force of real ones on the mind, they have manufactured consciences answering to the imagined realities, and so have invented a thousand duties that did not really exist. Carrying this thought out now by some illustrations from practical life, we shall be able clearly to mark the distinction between a true and a false conscience. A true conscience is one that answers to, and takes cognizance of, the real relationships in which we stand to God, the world about us, and our fellow-men. A false conscience is one that answers to, and is formed in accordance with, what the man supposes to be true, but which is really false. Let us take some specimens from real life.

I. A false conscience.

The early Greeks and Romans believed heartily in the supposed fact that Jupiter, Juno, Mars, Venus and their fellow gods and goddesses held a celestial court on the summit of Mount Olympus. To their thought these were real divinities, ruling the world, and holding in their hands the destinies of nations and of men. Their moral sense of right and wrong, their conscience of religious obligation, took hold of, and adjusted itself to, these supposed but really non-existent facts. Thus, out of what we know was only inherited superstition that had no objective reality, they constructed a whole religious system by which they were bound with the strictest sense of obligation. They built temples, and instituted rites, and felt their whole life touched and shaped by what to them was the real sense of the overlooking and ever-present gods. We know to-day that what they so laboriously performed were no real duties at all. Their fears were groundless, and their hopes were dreams. They had a vital conscience of what did not exist. These same people, believing that the proper burial of the body was intimately connected with the future rest of the soul, would suffer conscientious pangs for years on account of any neglect of the funeral rites. We know that all this fear was purely imaginary. I have spoken of tribes that held it a religious duty to put their parents to death before they grew old and infirm, because they believed that, if they entered the spirit-land withered or old or decrepit, they would remain so forever. Thus their consciences turned murder of father and mother into the most sacred religious duties. We, knowing that these supposed facts are only superstition, regard as horrible what they think religion. Paul, believing in Moses as against Christ, thought it his duty to persecute the early Christians. Afterward, when he changed his thought of the relations in which he stood, he found out that his first conscience in the matter was a false one. The first Christians, believing that Christ was coming immediately in the clouds of heaven to destroy and reconstruct the world, held it their duty to oppose property and marriage, and the whole order of things on which civilization depends. Their conscience in this matter was false because their supposed facts were untrue. The Romanist worships Mary, and prays to saints, and trusts in relics and superstitious ceremonies. His conscience binds him to this; but we know that his conscience is false, because it is made to accord with a whole world of imaginations that are no facts, and that do not at all touch the question of his moral goodness, or real relation to God or man. The early Puritans felt conscientiously bound to drive out the Quakers, and burn the witches; but theirs was a false conscience, because the Quakers were as sincere religionists as themselves, and there were no such things as real witches. Moody and Sankey believe in an angry God, and an everlasting hell; and their consciences are made to answer to these supposed facts; but, disbelieving what to them are vital truths, we conscientiously oppose their methods and their work as dishonoring to both God and humanity. So, all about us, we see men and women who have a false Sunday conscience, or business conscience, or reform conscience, or prohibitionist conscience, or political conscience; and these men will be conscientiously unjust and uncharitable and persecuting and cruel. The most uncomfortable man in the world to get along with is one who is pig-headedly obstinate in his conscientiousness. He is so certain that he is doing God's work, that he will be religiously inhuman about it. He gets so anxious to drive men into heaven, that he will use the spirit and weapons of hell in the work.

Now, all this false conscientiousness grows out of the fact that men suppose they stand in certain relationships that do not really exist. Thus the imagined duties are no duties at all.

II. A True Conscience.

The nature of a true conscience appears in the fact that it is the very opposite of the false. It is one that answers to the real facts and relations of life. I am under no real obligation to any imaginary God, whether he be the dream of a Hindoo or Greek or Jew or Christian. Toward the God of Pius IX., or of the Sultan of Turkey, or of the Emperor of China, or of the evangelist

Moody, I feel no gratitude, and admit no sense of duty or obligation. The real duty lies toward the real and true God who is manifested in the great laws, forces, and realities of the world, and in the highest and purest life and thought of humanity. Though each narrow religionist calls it infidelity and atheism, yet the truth of the true God is the only true religion of which the true conscience will be the echo. Orthodoxy regards the liberal as derelict in duty because he does not join him in his work of soul-saving. So does the Mohammedan look upon the Christian because he does not seek his heaven through Allah and his prophet. But the God, and the heaven, and the hell, and the lost humanity, of orthodoxy, we believe to be purely imaginary. Thus their all-engrossing work is not only no real duty to us, but it is even effort thrown away, power wasted, that might be rightly used in real help for man. A true conscience, then, is one that knows and is adjusted to the realities of life. When men know the truth about God, about themselves, — body and mind and spirit, — about the real relations of equity in which they stand to their fellow-men in state and church and society, and when they appreciate these, and adjust their consciences to them, then they will have a true conscience. An absolutely true conscience of course cannot exist so long as our knowledge of the reality of things is only partial. We can only make closer and closer approximations to the ultimate truth. So conscience will keep step with the progress of the race. Just as the old theories of the universe are being constantly corrected by the advancing knowledge of the world, and the real facts of things are coming to light more and more, so will the sense of duty grow, broaden, and advance, re-adjusting itself continually to the higher intelligence of the time. It is only beginning to be seen how conscience must have to do with the care of the body, with sanitary regulations of cities and homes, with the proper treatment of the criminal classes, and with the whole physical life of the world. "The thoughts of man are widened with the process of the suns." And so the conscience of man will widen to the measure of his thoughts.

But in the midst of this progress, and since the conscience is not infallible, which way lies present, practical duty? In the first place, I would say, Whatever any man's conscience commands, that he must do. It is to each soul the supreme voice of obligation; and he must obey it on penalty of violating and weakening and killing out his moral sense. Evolution will insist on this just as strongly as the old orthodoxy. Whatever you are persuaded is wrong, that you have no business to do. Whatever you think is right, you are under the highest obligation to perform. Even though to others it be only a whim or a prejudice, it is to you supreme duty. Whether it is a Romanist refusing meat on Friday, or a High-Churchman staying away from opera in Lent, or an Orthodox refusing to ride in a street-car on Sunday, still each one must obey the dictate of conscience. If the pope thinks Luther is sending men to hell, then he

must seek to crush him; and if Luther believes he is
standing for God, then he must defy the pope. And here,
in this clash and conflict of consciences, come the great
tragedies of history. Men and causes temporarily go
down; but God and truth ultimately win the day for a
higher humanity. Out of the tempest comes a clearer
air in which the outlines of truth are more distinctly
visible.

But, secondly, each man must remember that his conscience is not infallible, and that it may not represent the
real truth of things. So there comes in a duty that over-
tops all others, — the duty of perpetual, earnest search
for truth, that so the conscience may be made to accord
with the real facts of God. You must keep ever your
eyes open for light, and your heart open with a world-
wide welcome for whatever is pure and true. Remember
that we are finite minds in the midst of infinite realities.
"We know only in part;" we must seek to know as
completely as we may. Throwing away prejudice and
conceit, seek to make your conscience like the magnetic
needle. The needle ever and naturally seeks the unchanging pole. Ignorance and false teaching, and self-
confidence and passion, are side attractions that deflect
it from the right. Surround it only with truth, and it will
guide the way to God and the eternal life. "A good
conscience" is to be found only by finding out the
real truth of things, — truth that can be verified and
settled as true, — and then by a constant and sincere
obedience. With loyal heart, then, anxious to do the

right when discovered, let our daily longing and prayer be the sincere cry that the Grecian Ajax sent up out of the darkness of his conflict before the walls of ancient Troy, — *for light*.

> " Walk in the light : so shalt thou know
> That fellowship of love
> His spirit only can bestow
> Who reigns in light above.

> " Walk in the light : and thou shalt own
> Thy darkness passed away,
> Because that light on thee hath shone
> In which is perfect day."

VII.

LOVE IN LAW.

Law, in the popular mind, means all that is cold, hard, heartless, and cruel. This feeling has been fostered somewhat by inconsiderate writing on the part of some scientists, or by popular misconception of scientific writings. But the larger cause has been the tone of theology and the pulpit. These have represented law as only police and hangmen, thunders of Sinai, or torments of hell. For ages the plan has been systematically pursued, of making men feel that they were in the clutches of cruel law bent on their destruction, and then of offering them deliverance from its iron power by means of a gospel of mercy and love that was represented as being outside of and above law. Thus the belief has been created, that law and love were of necessity in a kind of eternal antagonism. To the question, then, of the nature and results of law, whether it is loving or cruel, we will now address ourselves.

The early conception of the universe was that its great forces were predominantly cruel. Cold, darkness, storm, hunger, wild beasts, death, all these were the work of evil beings that had no love for man, and were instigated by hatred and malice. Nature was cruel. It was not, however, cruel law. These early men had no conception of any such thing as law. Cause and effect had no orderly relation in their mind. It was only cruel caprice. Behind these great movements and forces were malignant persons that loved the smell of blood, and rejoiced in the infernal music of human groans, and eagerly licked up the tears that sorrow let fall. This is true of all the early religions. The popular conception of Jehovah among the Jews was that he was hard and cruel, the god of thunderbolts and hail, "a consuming fire," one punishing children for the sins of the fathers. At first, then, nature was cruel caprice.

In after-times there gradually grew up a belief in a good god. But you must notice that he was not at all the god of these supposed cruel forces of nature. They were bad gods still in all their terrible reality. Only another god appeared, who was good and kindly disposed toward humanity, and who opposed and sought to destroy the works of the wicked nature-deities. This good divinity was outside of and wholly separate from nature, so that nature was still regarded as evil. This belief was fully developed in Persia, where Zoroaster had set in eternal opposition the good Ahura-Mazda and the evil Ahriman. A similar belief appears in early Greece, where Prometheus is chained forever to the rock, while a vulture eternally devours his vitals; and all because he had shown himself disposed to help mankind by revealing some of the celestial secrets. He is the good god overpowered by a majority of the Olympians, who were evil disposed toward men. A similar idea appears in early Christianity. God the Father, laying aside the Old-Testament hardness, appears as one who loves and wishes to save humanity. But the world of material forms and forces is all opposed to him, being under the dominion of the "Prince of this world." God in Christ was to deliver from his power, and from the domination of the world of sin and death, all who put their trust in him, and renounced loyalty to the devil. Christendom has generally stood for this phase of belief. There was not yet any law, in the modern sense of that term: it was only that God stood apart from and opposed to the evil forces and dominion of nature, ready to save out of it any who would accept the conditions. From this has sprung up the present popular conception of nature and law as something apart from and opposed to God, so that the religious literature of the time is full of the supposed need of a loving power to care for, protect in, and deliver from, these great natural laws and powers. The thought is, These laws are hard and cold and cruel: they do not heed our tears ; they do not care for our prayers: we want to believe in some one outside of and above law, who will turn aside its wheels, and save us from being crushed under the onward roll of this Juggernaut.

But in very modern times, and in spite of all sentiments of fear or feeling, wide and deep study has developed a knowledge and sense of law, as opposed to what is regarded as the capricious action of will. And this belief in law threatens to cover, with its network of cause and effect, all the universe. Science recognizes everywhere in nature the dominion of law. It has also demonstrated the law of development, or evolution, that brings into the general order the whole movement and upward march of life on the globe. It talks now of laws of history, laws of society, laws of virtue, laws of crime. The most changeable and apparently capricious movements and forces, it threatens to catch and hold in the meshes of its universal order. And all the while, in the popular mind, this law seems to exclude God and his love, and to leave all things in the hands of a hard and heedless fate.

Men wish to know, then, whether this is true. If law is everywhere, is love shut out? The popular conception of love, as something outside of, above, and delivering from law, is shut out. God himself is not above law; for law is order and reason and the cosmos. Put God outside of law, and you make him disorder and unreason and chaos. Not that God is subject to any law higher and stronger than himself; but that he must live out the law of his own being. That man is subject to the law of his own being, is only to say that he is sane. Superiority to law, order, and reason, is what we mean by insanity. The condition, then, must be one of two, — law without love, or else love in law; law the expression of love; for love without law, or above it, or outside of it, is only unreason and chaos. Love in law, then, is the thought of the highest and clearest thinkers; and it must be the ruling thought of the future.

These three stages of thought the human mind has passed through in its upward evolution. First, cruelty as caprice; next, love as caprice opposed to capricious cruelty; and, lastly, law,— first thought of as hard and heartless, and then lifting up into all-dominant love, as itself of necessity law and order, superior to caprice, and, as being perfect, incapable of change. This last is the position of "the religion of evolution." Let us now see if we can comprehend something of its truth and beauty.

And, first, we will see how broad and all-inclusive is this universal "reign of law." As we have seen, the early races put persons everywhere. There was a god of the year. Day and Night were deities. The Dawn, a goddess, led forth the dance of the rosy-fingered Hours, twelve other deities. The Sun, a god, drove his flashing chariot-wheels across the solid roadway of the sky; and at night the Moon, "pale goddess," ruled the dusky Hours, and led out the Stars for their night-long choral song. The wind, and the clouds, and the light, and the sky, and the rivers, and the seas, and the trees all had their god or goddess. Nothing ever came to pass except as the work of some personal will. And these wills had no conference or understanding together; but each did "what was right in his own eyes." The god of the winds made them blow to suit his own whim or caprice, or to help or harm, as he took a notion. The god of the sky sent rain or dew or bright days, as best accorded with his own convenience, or as moved thereto by prayers or sacrifices. The god of the grapes gave plenty of wine, or blasted the ripening harvest, as he pleased. Ceres blessed the corn, or cut off the crop. Nothing was supposed to occur according to fixed and calculable order. All was practically chance, because no one knew beforehand — except by oracle or prophecy — what any particular god might conclude to do. This was the condition of thought throughout the so-called heathen world.

Among the Hebrews it was very much the same. While, in their later centuries, they were monotheistic, and recognized but one God, the supreme Jehovah, they yet thought of all natural forces as under the direct and constant superintendence of angels, who ruled as vicegerents, or satraps, over the different departments of the world. There was an angel in the sun, an angel in the moon, angels in the stars, angels of the winds and waters, rivers and trees, angels of rain and storm and darkness, angels of the growing crops and of flowers, angels everywhere, and the active agents in every thing. This system differed from the "heathen" only in this, that these angels were supposed to be subject and answerable to Jehovah, who was the only king. But this difference was partly broken down by the belief that sometimes these angels were rebellious and faithless to God, and ruled wickedly after their own evil thought and will.

Thus, at first, law was nowhere. This in general was

the condition of affairs until very recent times. Perhaps few are accustomed to think how modern a thing is this conception of law. Even in the time of Kepler, the wisest astronomers must think of an angel in the sun and moon and stars to account for their movements. The first grand triumph of law was in Newton's discovery of gravitation. This bound the orbs of the material universe in the perfect and beautiful order that to-day makes astronomy the most exact and fascinating of sciences. But by many the comet was still regarded as an exception, a lawless and unaccountable wanderer up and down the heavens. But when it was found that gravitation was able to calculate even his apparent irregularity, and to tell, from the path of his departure, just the year and hour of his return, even over the gulf of centuries, then the comet himself began to keep time and step with the orderly march of the "host of heaven," the armies of God. From that day to this, the advance of knowledge, and the discovery of law, have kept even pace. Caprice and miracle and spirits are still sometimes called in to explain what else is not understood. But we know from the past that a thousand things considered miraculous and inexplicable by natural law have at last been explained, and are now thoroughly understood, so that from known causes certain and definite results can be easily and always predicted with an absolute certainty; and, if any thing can be learned from the logic of the past, it is certain that the mysteries and unaccountable phenomena of the world will one day be known, and reduced to the natural and universal order of cause and effect. Caprice and lawlessness are already driven into the dark corners of the earth, where the light of knowledge has not yet penetrated. But this is without exception true, that there is absolutely nothing that is known which is not subject to and explainable by law. Let us look about us a little and see, so as to impress upon ourselves this truth.

Every sun and moon and star and comet is held in and guided by the reins of an omnipresent law. Even their perturbations and irregularities can be accounted for and predicted beforehand. The development of the earth, from its molten to its habitable condition, has been in the hands and under the guidance of a law so perfect that the most remote and various results might have been seen and predicted from the beginning. The history of the rise, the growth, and the decay of nations, also is only another illustration of law. The nature of their origin, their elements, and their surroundings, have determined their careers and the "bounds of their habitation." It matters not that they themselves have not recognized it. The drops of water in a stream do not recognize the currents and the banks; but one standing on the shore can trace and formulate the law of its whole movement from source to mouth. So, as we study the past of humanity, we can see that the movements, the wars, the conquests and defeats, the overturnings, were not matters of caprice or chance, but were governed by all-determining laws. So true is this, that one who studies France, or Spain, or Germany, or the United States, can trace the order of its destiny, and predict the main outlines of its near future. Even the fair forms below us, the slightest and most frail, tell the same story. The wild field-flower, in all its freedom of growth and grace of development, is the handiwork of law. Down in the little earth-hid seed, and in the invisible forces of the air, are the powers that control its apparently wanton growth, and unroll its tinted petals until they look up perfect to the sun. It is the product of conditions so fixed that the size and shape of its stem, the number of its leaves, the shading of its colors, and the quality of its odors, could not possibly have been other than exactly what they are. And so far is this from marring its grace and beauty, that it requires the perfect laws to produce the perfect finish. Thus the nice hand of law has wrought for us the million varieties of the grasses and the flowers; and they are what they are simply because they came up in "the law of the Lord."

The same is true of those forces of nature that seem the most lawless and uncontrolled. Even the wind, as it "bloweth where it listeth," always "listeth" to blow in the way appointed it. It has its fixed courses, its going out and its return. It runs swiftly in well-defined channels, as truly as do the rivers in their banks. And while seemingly loitering on, toying with a flower, or a lock of hair, or fanning an invalid's cheek, its way is as sure as that of the eddying ripple on the brook's surface. And when the sweeping hurricane drags on through the wild heavens its black car of desolation, it rushes on its appointed way as really as do mountain torrents that the rains have swollen. So the lightning strikes no uncertain blow; neither do the surcharged clouds of the tempest drift aimlessly in the sky. Law sits as charioteer, and holds the certain rein. Free, wild, unchanged, the fury of the storm is as truly in the hands of law as is the electric flash when bound to the continent-spanning or ocean-diving wire, and bearing regular messages for man. Gen. Myers sits in Washington, and, reviewing the winds and the clouds all over the continent, is able to tell us day by day, whether or not we shall need our overcoats and umbrellas.

It is law, too, that gives us the order of the seasons. Its mighty arm swings us through the circuit of the zodiac, and fills our hearts and lives with the variety of the year. In this confidence we know that "summer and winter, seedtime and harvest, shall not fail."

So in our individual life. There is no part of our being that law does not touch, mould, and direct. By law we make every motion of our bodies; by it we breathe, and enliven and color the blood; by it the blood courses vein and artery, repairing nature's waste and wear; by it the heart beats, and the brain thinks, the eye drinks in beauty and the ear, sounds; it compasses us round, and hedges us in on every side; but do we feel it a bondage? Not if we are healthy. The least lawlessness is incipient disease; more becomes insanity or death: so law is life.

And all this is just as true of thought and feeling,

the intellectual and the spiritual realms. Law and order are everywhere the conditions of life, of happiness, of goodness, and of beauty. So wide is its sweep, and so grand is its unity, that Herbert Spencer has been able to think and to formulate in a few sentences a *law of the universe*, of which all things are only illustrations. This is the grandest achievement of the intellect of man.

Now, whether we like it or not, it is pretty well settled that we shall have to accept this universality of law as a fact. All the knowledge of the world points to it more and more definitely as true. Law and order reign. Curses do not bring storms, nor prayers avert them. Every department of the universe has its own laws and conditions; and events are controlled solely by these. Not only is this so, but the highest religious conception of God is coming to demand just such a state of things. A God who constructs a system, and then has to keep coming in to regulate and re-arrange its working; who changes his mind, and fixes things over after another and a new plan; who could not see and order all things from the beginning; who submits to having his elbow jogged, and to take wiser suggestions from men than any he had before thought of; who comes to help out of difficulties that his own foresight might have provided against; who has to keep mending and looking after things in an abnormal way, — such a God, it is beginning to be seen, is only a being created in the image of man, and in the image of a man who is not over-wise or over-good, either. To him who has gained a higher conception of God, a conception so grand as to be in keeping with universal and eternal law and order, it seems impiety and irreverence to think of him after any lower fashion; just as to the ordinary Christian it would seem impiety to represent his God as capable of being pleased with the groans and the burning of a human sacrifice, or as being liable to human ignorance. So any thought of God less high and grand than that which embraces universal and eternal law and order seems irreverent and degrading.

But now let us see if there is any place for love in such a system. I am ready to admit that the heart cries out for love just as loudly as the brain calls for law; and, further, I am ready to admit that to gain order for the head at the price of the loss of happiness and trust for the heart is a most questionable advantage, or even a positive loss; for the heart and its needs are as real and true and high a part of human life, as is the knowledge and thought of the brain. I even believe that happiness and peace are so necessary a part of life, that any life is a failure that, in the long-run, does not gain them. If I would hesitate to say, "Where ignorance is bliss, 'tis folly to be wise," I would not hesitate to say that I would oppose knowledge, did I not believe that its outcome would result in the highest and truest welfare and happiness of man. So I would not slight the heart and its affections, even for the sake of the grandest truths and laws.

Instead, then, of doubting about the answer to the question as to whether there is any place for love in a system of universal law, I am ready to make the claim that there is no place in it for any thing but love. It is all one complete, perfect expression of the wisest and highest love. Consider, —

(1) That we know love and joy as facts of science. They reach from the depth to the height, and sweep through the whole length and breadth of animate life on the globe. The whole natural and healthful creation is one scene of sentient gladness. The fish balanced in his crystal home, or darting here and there like a beam of light, is full to his utmost capacity of gladness. Healthy life itself is an ecstasy. The insects that glitter and buzz in the sun are as full of joy as a liquid drop is full of water. And in June

> "The little bird sits at his door in the sun,
> A-tilt like a blossom among the leaves,
> And lets his illumined being o'errun
> With the deluge of summer it receives."

And so all through human life, just so far as there is health, harmony, and right adjustment between man and his conditions, there is a music of joy, like that of an instrument when kept in tune. And from the mother-bird in her nest, or the father-bird gathering food for the young, clear up through every grade, till you sit beside the human mother rocking her cradled babe, or watching over the restless sleep of some sick darling, there is everywhere the all-pervading and ever-growing love and tenderness that are the glory of the highest humanity. Now, all this love and tenderness are facts of life; and as such they are a part of the infinite manifestation of God. They have a source as much as do mountains and fossil bones. And we may justly reason that the fountain is more than all the streams. For the growth of civilization is the growth of love and care and help. Since, then, these are growing still, with no sign of diminution, we may believe that the source is so much greater than all we see as to justify us in asserting that all human love and tenderness and care are only faint glimmers and reflections of the divine. Your tenderest mother-love is no more when compared to the divine love than the shimmer of moonbeams on the water is equal to the glory of the noonday sun.

Law, or no law: so much we may say we know. To see, then, whether this fact is contradicted or overbalanced by the fact of universal law, let us go on, —

(2) To inquire, What is law? On this point there is a strange and wide-spread misconception. Theology has so long taught that nature was opposed to, or at least outside of and separate from, God, that it is a long and difficult task to correct the error. Natural law, then, is only God's method of working. The highest science teaches that matter and force and law are only manifestations of God to human consciousness. This is the highest wisdom of all the ages. Since God is equally everywhere, and in all things, science knows no distinction of natural and supernatural, or of strange things from one side breaking through and interfering with the regular

on-going of affairs on the other side. All law, then, everywhere is natural; and this natural law is simply God working. And even if we could not observe it all about us, the uniformity of his working would be a necessary inference from his perfection. To speak after human analogies, if, the first time God ever did any particular thing, he did it the best way, then, under similar circumstances, he must always do it the same way: otherwise he must do it in some poorer way, and that would be a display of imperfection and unwisdom destructive of our very idea of God. God's perfection, then, demands universal and eternal law as its natural expression. When we wish for a break in this perfect system, we are wishing for the dethronement or the degradation of God. But, —

(3) Take notice of the character of the working of natural law. It is always and everywhere for good. There is no one single natural law in all the universe, so far as we know, whose normal working is not for good. Why, then, would we have them weakened, broken in upon, or changed? All evil is only law broken or disobeyed. Even pain is only a signal marked "Danger!" that is set up along the railways, at the switches and crossings of human life. If it were not painful for the moths to get singed in the gas-flame, they would burn themselves up in its luring brightness. If pain were not the result of breaking the laws of the body, who would be careful to keep in health? Were it not for the pains of discontent in low conditions, what force would have

driven man on and up into civilization? It is just this perpetual unrest and sorrow in conditions of incompleteness and maladjustment, that compel the perpetual struggle for higher and still higher forms of life. It is a serious question as to whether any useless pain can anywhere be found. Certain it is that the natural, healthful working of all the forces of the world issue in goodness and joy. Health and harmony and obedience are everywhere music. Pain is simply God saying, "Get out of danger," or, "Go up higher."

A fearful picture is sometimes drawn of the battles, the rapine, the blood, the mutual devouring, of the animal world; and the argument for nature's cruelty is drawn from it. It appears to me to be one huge fallacy. We carry our human sympathies and hopes and fears down into this lower life, and judge it by our standard. Look at the facts. Animal life in the main, and all through, is happy: they are content and satisfied. But death is a part of the law of life: so all must die. This is no wrong. A good that I keep for a year is not an evil, because it is then withdrawn. Now, it is doubtless true that, in being slain and devoured, the lower animals suffer less than they would by lingering, and dying a natural death, perhaps from starvation. They suffer from fear so long as this fear can aid their escape; but it is well known that, in almost all cases when the prey is caught, fear is gone, paralysis takes its place, and death is painless. Dr. Livingstone relates that, when the lion's paw was on him, he was stunned, and had lost all fear; and a similar thing is true of all the victims of carnivorous birds and beasts. Natural law, then, everywhere works for good; and until some law is discovered whose normal working is evil, the statement that law may justly be called an expression of love must stand unimpeached.

Law, then, makes the order, beauty, and music of the heavens. This is the true "music of the spheres." It is only law that makes civilization possible. It is because we know the laws of earth and iron are constant, that we can build railroads; and because the laws of steam are changeless, we can run our engines upon them. With faith in the unchanging laws of wind and wave, we build and sail our ships all over the globe. On the laws of gas, we build and arrange the lamps that turn our streets from darkness into illumination. The law of electricity assures us that the cable that to-day takes our message to London will not be useless to-morrow. By the laws of nature we erect the stores and houses, and lay out the streets, of our cities. It is the stability of God in our bricks and stones and timbers, that makes us rest at night, with no fear that he will change his mind before morning, and let the whole thing down about our ears. On our faith in the laws of light, we put the plate-glass in our windows, and cut our jewels. Because laws change not, we have to-day the pictures and statues of the masters of the olden time. It is the stability of the laws of human life that makes it possible to frame constitutions and establish governments. If, in the long-run, humanity were capricious, statesmanship would be impossible. Art and science would be unknown but for law, that holds all things fast and firm. Commerce means trust in the general stability of human nature. Crimes and betrayals are only insignificant exceptions. That Shakspere and Homer hold their place in human thought, is only because the law of human thought is uniform. There could be no growth or development of religion, were it not for the constant elements of human thinking and human conduct. Were there no law in character, if it were the caprice of a self-moved will, how could we ever trust each other? That a man had perilled his life for virtue to-day, would be no guaranty that he might not commit murder to-morrow. That there is a cosmos at all, a universe in which life and thought and knowledge and progress are possible, is just because of the universality of law.

I do not wish any gospel, then, to deliver me from the law. To deliver my heart from the law would be to make me capricious, as likely to hate as love. To deliver my brain from the law would simply mean insanity. To deliver my body from the law would be disease or death. Whatever lives, lives in, and whatever grows, grows by, law. Happiness and heaven were delusions without it. Law, then, is the all-present, wise, loving God. God's law comes in the light, and wakes me for my morning labor. By his law I remember yesterday, and link the past with the present. By his law I break my nightly fast, and fit my body for duty and joy. By his law I live and think and labor and play. By his law I build my house, and do my work. By his law the world is full of life and brightness and opportunity. By his law the languor of sleep once more comes over me, while still his law makes my bed, and rocks me to rest on the old earth's bosom, "as she dances about the sun."

Do we, then, need any providence as a protection against or deliverance from nature and law? That could only mean that we need to have God defend us against himself. God is in law, and law is in God. This universal law is only universal, all-encompassing, tireless, changeless providence. It hurts us only when we transgress. But, then, it is for our good to be hurt, and so warned, lifted up, and delivered. "The law of the Lord," then, "is perfect," perfect wisdom, perfect power, and perfect love. May we not, then, on our part, exclaim with the old-time Psalmist, only with fuller meaning, "O, how love I thy law! it is my meditation all the day"? And may we not fitly close with the upward-reaching cry, "Teach me, O Lord, the way of thy statutes, and I shall keep it unto the end"? For "the law is holy and just and *good*."

VIII.

PRAYER.

The place which prayer holds in the popular thought, the controversy in regard to it, and the consequent doubt and confusion, — these make it a topic of the first importance both to the understanding and the heart. To glance at its origin and development, and then to study the popular beliefs in the light of modern knowledge, and so, if possible, to gain a position both rational and devout, — this is my present purpose.

I have shown that the early objects of human worship were not persons to be loved. The gods of the childhood of the world were beings who filled the souls of men with dread. The first altars were built by fear; and the first prayers were deprecations, pleas, and entreaties intended to ward off impending ill. It was common among the early tribes to worship the spirit of their ancestor, who was supposed to watch over his descendants, and to hold their fortunes in his hand. This worship did not spring from any special love or veneration; but they believed that their tribal founder was jealous of his memory and honor, and that, if they did not keep alive his name and his worship, he had both the power and the disposition to do them incalculable harm. So their altars and their prayers were only a sort of tribute paid to keep their god in good humor, and thus to buy his favor and protection.

And a similar idea prompted the aboriginal worship of the forces of nature. The forces and movements that pressed most closely upon them were those of hunger and cold and storm; forces that hurt them, and of which they stood in continual fear. Thus their prayers were not at all in the nature of loving communion, as of a child with father or mother. They did not love their gods, nor care for their friendship or fellowship for its own sake. They only stood in dread of what the gods might do, and thought it the safest way to keep on as good terms with them as possible. So their worship took on a sort of commercial aspect. They thought to buy favors, or at least freedom from injury, as one would put a bribe into the hands of an unjust judge, or purchase with a gift the good-will of an irresponsible despot. When they went out to war against their enemy, they first sought the altars or temples of their god, endeavored to learn his disposition in the matter, appealed to his ambition as against the honor that the enemies' god might gain if he did not show himself the stronger; they promised special worship and offerings if they came back victor; and then, after the battle, they came and hung their trophies in his temple, and ascribed to him the praise of their conquest. You find ideas like these even among the Hebrews. Moses is represented as appealing to the pride and ambition of Jehovah. Weary with the obstinacy and obtuseness of Israel, Jehovah threatens to destroy them. Then Moses pleaded with him, and said, "Only think how Egypt and her gods will exult, if, after leading thy people into the wilderness, thou leavest them to perish, as though thou wast not able to lead them on to the conquest of the Promised Land." And the plea prevailed, and the people were spared.

The old-time gods were accustomed to pray to each other; for as each had his department of the world, if he wished to gain any end beyond his control, he must do it through the favor of the deity in command. Æneas and his followers are on their way from Troy to Italy, when Juno, who is their enemy and wishes their destruction, goes to Æolus, the god of the winds, and by a condescending appeal to his pride and friendship, and by the promise of a magnificent gift, she persuades him to let loose his winds, and raise a tempest on the sea, so that the Trojan fleet may be destroyed. At the same time, Venus goes to Jupiter, and begs for their deliverance; and the king of the Olympians sends a messenger to see that the storm is allayed, and the ships are permitted to find a harbor of safety.

Such were the early ideas of the place and the work of prayer.

All through mediæval Christianity, similar ideas prevailed. Each city or people or convent had its special tutelary saint; and to him, as to a favorite at court, their petitions were addressed. He was supposed to have influence in heaven, and to look after their peculiar and personal wants. God, to them, was not the all-present, watchful Father. They thought they needed special intercessors and friends at the heavenly court. And even to-day, throughout Romish Christendom, the great majority of prayers go up to the saints and to Mary, whose tender mother heart is supposed to be most easily touched and moved at their requests. And a thought like this last seems to be infecting all Protestantism. The Bible nowhere gives countenance to making Jesus the object of prayer; and most intelligent and thoughtful writers oppose it as unscriptural and wrong: but in the popular mind the Father has come very largely to represent the law, while Jesus is the embodiment of the loving and saving God; and thus to him the tenderness and love of all hearts turn, and to him are addressed the pleadings and the prayers of the anxious and fearful souls.

The common idea of prayer to-day is, in the main, that of the ancient heathen world. As I have many times heard it from the lips of a prominent evangelist, "Prayer is the power that moves the arm that moves the world." It is supposed that by it God can be prevailed on to do many things he otherwise would not. Stillingfleet, an old English writer, says, "Prayer is supposed a means to change the person to whom we pray." It is believed to have power to bring to pass the things prayed for all over the world.

Within a few years, in Scotland, the whole Church was ordered to hold a service of fasting and prayer for deliverance from a severe disease among the cattle. Only a year or two since, the whole Church of England was holding a special service for the recovery of the Prince of Wales, who was sick. Müller, in his "Life of Trust," claims that he has established, and that he constantly supports, an immense orphan-house simply by prayer. He seems entirely to overlook the fact that such a claim, continually republished and kept before the popular mind, is the most ingenious and effective kind of advertising possible. It is a perpetual appeal and prayer to the popular sympathy and help. To one who at all appreciates the springs that move the popular heart and the popular imagination, there is nothing strange about it. Dr. Cullis, of Boston, makes a similar claim concerning his Consumptives' Home; and yet hardly an institution in the city is more persistently advertised. He also has his troops of children travelling all over the State, visiting the churches, and exciting the popular interest in the piety and the wonder of the work he is doing. In one of his books of sermons, Mr. Talmage ascribes the remarkable success and safety of a certain line of ocean-steamers to the fact that the wife of one of the head proprietors makes each ship, as it starts from port, the subject of special prayer. This theory is slightly interfered with by the knowledge that another line, that has no proprietor's wife to pray for it, has lost less ships in thirty years than it has. And, besides, were this all true, insurance-companies would be an impertinence, as well as a useless outlay of money. Most of our churches in America have ceased praying for rain; though still the annual fast is appointed, and is supposed to have some influence on the general welfare of the people.

Now, all these illustrations simply point out and illustrate the popular belief that prayer is able to induce God to produce certain definite results in the natural world that would not take place except for the prayers. You will notice, also, that this whole conception of the nature and office of prayer implies that God is a person outside of and separate from natural forces and laws, who, at the request of man, comes in to interfere with and change the method of their regular working.

Along with the progress of civilization, and the growth of scientific knowledge of the world, there is a growing a wide-spread doubt as to whether prayer has any such power, or produces any such results. Whether for good or evil, whether well founded or ill founded, this doubt is a fact. It is a doubt that touches and paralyzes the arm of the Church itself. The ordinary prayer-meeting is not attended in any such way as indicates a vital belief that it has power to convert the heathen, to save the husbands, brothers, fathers, of its supporters. People go from a sense of duty, or as they feel inclined; and if it is made a sort of lecture, and the lecturer is interesting, then the room will be full. But if men believed that it was really a power to move God to their wills, how could they excuse their negligence? The constant appeal and devices of the minister, to get even church-members to come, show how little real faith they have in it.

Let us, then, inquire whence this doubt springs, and whether it can be justified. There are certain great difficulties in the way of holding the traditional faith that press very hard on thoughtful minds.

(1) Prayers for changes in the natural course of things, so far as we can ordinarily see, are not answered. This is so generally admitted, that I have often heard ministers tell their people that they prayed as though they did not expect to get any thing; and that probably nothing would be so much of a surprise and astonishment to them as to find their requests granted. Now, people in that state of mind cannot have received what they asked for, very often. And, for my part, I cannot be particularly surprised that they do give up looking for it. "I have been praying for the conversion of such a one twenty years, and he is not converted yet," said a noted minister. Is it strange that people ask, "What, then, is the use of prayer?" But it is said, more prayers are not answered, because people do not pray in faith. So the mediums tell us that the séance is not successful, because sceptics are in the circle. But if the spirits are in earnest, why do they not astonish and confute the sceptics by offering facts that would compel belief? and if God really wishes to answer human prayers, in the ordinary sense, why does he not astonish his listless children with answers, so that they cannot help believing?

It is this last rational thought that gave life to the

recent celebrated "prayer-gauge" controversy in England. And if prayer is a real and definite power for the healing of the sick, I see nothing irrational or impious in the test that was proposed. In the Old Testament, God is represented as complying with man-suggested conditions so as to demonstrate his presence and power. Has he changed, so that he is not willing to do it now?

A few days since, the newspapers reported the miraculous cure of a sick woman as the result of Dr. Cullis's prayer. If true, why is not the road to Grove Hall thronged with the lame, blind, and diseased, going to be healed? Why does he keep his "Home" at all, at an enormous expense that might be used in other beneficent ways? Cure them all, and then travel the country as the universal healer. I have it from a man who was a relative, that Joe Smith, the Mormon prophet, healed a child of a chronic disease, by the laying-on of hands. There is evidence to prove that Our Lady of Lourdes is continually curing her devotees. And a large mass of proof can be brought to show how the bones of saints and bits of the true cross have always exerted such power.

The simple facts are these: Talmage prays for his wife, and she gets well. Another man, presumably as pious as he, prays for his wife, and she dies. A third man's wife gets well although she is not prayed for. The laws of health and disease work all over the world, and produce their proper and natural results; and, so far as we can settle by the evidence, the praying or not praying has nothing to do with it. Gen. Butler goes to New Orleans;

and the saints of the region pray that "Yellow Jack," as they call the prevalent fever, may waste away and destroy his army. Not being given to piety, he does not pray against it; but having a sense of natural cause and effect, he looks after the sanitary condition of the city; and "Yellow Jack" postpones his annual visitation.

(2) Superadded to this doubt growing out of observed facts, is the essentially modern doubt that has its root in the growing knowledge of the reign of law. There is growing up an irresistible belief, based on facts that can be verified, that this universe is one ruled by law and order. Cause and effect are so intimately linked together that the slightest movement of the world's affairs to-day is a link in a chain that stretches back into a past eternity. The glance of a ray of this morning's rising sun, the tremble of a twig in the wind, the form of the smoke-wreath that hovered above your chimney and then melted out of sight, the curl of a wreath of mist that floats on a mountain's top, the eddy of the little cloud of dust that the wind-gust whirls across the street, — these all are as much a part of the fixed and determined order of the world as Mont Blanc is a fixed and definite peak in the Alps; and to change the mist-form or the wind-current were as much a miracle as to hurl Mont Blanc with its roots upward into the Mediterranean. The prayer that asks that this day's weather may be changed, even to the blotting out of one cloud, or the adding or taking away one rain-drop, asks so stupendous a thing as that the whole order of creation from the beginning

may be changed to suit the whim or convenience of an hour.

(3) Then, again, the common belief is rendered absurd by the contradictory prayers of men. You pray for a bright day, that your excursion to the country may be pleasant, or that your hay may dry in the field. Another man prays for rain, that his parched potato-field may be revived. Which will God hear? One prayer at least must be denied. Two ships are at sea. One wishes an east wind, and the other a west. Others want neither east nor west. Dr. Bushnell once suggested that the majority prayer would win. A very large and very pious crew would seem to be a necessity, then, so that they might out-pray the other ship. When you think of it in this way, it comes to look as though these prayers, instead of being piety, might very easily come to be the quintessence of selfishness. Is it not quite as well to leave God to look after the general good, and not think to tease him into the favoritism of neglecting the common interest, that may demand dry weather, for the sake of watering your flower-garden? Heaven has larger interests on hand than the making of itself your private watering-pot. It looks very much like the little boy who wanted the captain to stop the ship, because he had lost his apple overboard.

But a stronger objection than all these roots in piety itself, and gets its weightest reason in reverence and trust. If God is indeed perfectly powerful, wise, and good, it becomes the worst of all scepticism to suppose that God

will neglect the smallest interest of any, the very least, of
his creatures. In the hands of such a God, the poorest
worm and the grandest archangel must be alike safe.
There is a mother in Boston who devotedly loves her
children. She is exceptionally intelligent, and knows
what is for their truest welfare. She is wealthy, and able
to command all means for carrying out her purposes.
One of her children is wayward, and gives her anxiety
and trouble. Another is sick and in danger of death.
Suppose you should go to her each morning, and with
tears in your eyes and sobs on your lips beseech and
entreat her to do her plain and simple duty by these children, — beg her to be kind and helpful to the wayward
boy; beg her to have a physician and nurse for the sick
girl. Such action on your part would be gross and inex-
cusable insult; and, if you repeated it, she would be justified in calling the police for your benefit. And yet half
of our prayers imply that, unless we keep jogging his
memory, or stirring up his benevolence and pity, the tender and all-wise and loving Father in heaven will either
not remember, or will not care, to be decently kind and
regardful of the welfare of his children. I have often
heard prayers, that people thought were pious, that made
me indignant, and that seemed to me simple blasphemy.
A hundred times, as a boy, I have heard a man say in
prayer, "It is time for thee, O Lord, to work!" — as
though God did not know whether it was time, or not!
Said another, "O God, *won't* you do any thing about it?"
Said an evangelist, "I am going to be in this city only

two or three days longer, and, O Lord, if thou art going to do any thing for the salvation of this town, do it now!"—egotism and blasphemy combined, with a doubt as to which predominates. Said the same evangelist, "Come to the meeting to-night, for there are going to be wonderful displays of the divine power,"—as though he had served a writ on the Almighty, and was going to produce and display him at all hazards!

Just because God is loving and wise and mighty, this kind of petition is not only useless, but insulting. And I make the charge that the prominent evangelism of the day, in its frantic appeal to God, as though it could not trust him to do right, displays a worse and more open infidelity than any it lays at the door of liberalism or of science. The quiet, loving life of the child at home is stronger proof of trust in father and mother than that anxiety that appears to think its wants will not be attended to unless as the result of perpetual begging. The little child in mother's arms looks up and smiles, and sinks off to sleep. It does not need to beg mother to rock its slumber, and tuck it into its soft crib. Just this child-relation toward God I believe to be the true and pious one. Prattle your childish wants in the Father's ear as much as you will: only remember they are childish, and that he knows best, and that the best of all prayers, after all, is, "Not as I will, but as thou wilt."

All this it has seemed needful to say before coming to that which is most important of all. The whole question finds simple, natural, and satisfactory solution when looked at in the light of the principles of evolution. What is that principle? This: All the myriad forms, forces, movements, and life of the universe, are only the varied manifestation of the divine life that lives in and works through it all. The divisions between natural and supernatural, sacred and secular, are broken down. All is natural, and all is divine and sacred. God is as much in the law of gravitation as in the moral law, "Thou shalt love thy neighbor as thyself." The sun, the winds, the clouds, the rivers, the growing fields, all speak his mysterious name, and reveal his present life and power. The law of the unfolding flower is just as divine as is the impulse to gratitude or prayer. It is just as rational and religious to expect prayer to yield to gravitation, as it is to expect gravitation to give way to prayer. God does not undo with one hand what he is doing with the other. Nature is no longer — as was once universally, and is still too commonly, supposed — a realm opposed to or outside of God, into which he comes at times to counteract its forces, or to deliver from its power. Nature is God at work, executing his own wise and perfect will. It is not to be supposed, then, that the impulsive requests of men are going to persuade him to contravene his own purposes, or interfere with his own wise work. It becomes not simply unreasonable, but a lack of intelligent piety, to expect it.

Do I take the position, then, that man can bear no active part in the midst of the divine operations of the world; that he is only to sit still, and let the great forces drift him on their current, like a leaf on the surface of a torrent? By no manner of means. Man's whole life, physical, intellectual, and spiritual, depends on just how much and how wisely he "interferes" with God's working, bringing to pass results that else would not occur. But he must do this intelligently, and according to the laws of that department in which he expects his success.

To illustrate what I mean: You are perfectly aware of the practical use of my principle in the common affairs of life. You do not abdicate your reason here. If you wish to move a piece of stone or timber, you use a lever or some mechanical force: you never think you can argue it, or pray it, or wish it, or will it, from one place to another: you meet mechanical force with mechanical force. If you desire to gain admission to a man's favor, you do not think you can pry his heart open with a handspike, or blow it open with nitro-glycerine: you meet emotion with emotion. If you wish to convince a man's intellect of a doctrine in political economy, you never think of applying electricity or steam: you meet logic with logic, argument with argument, proof with proof. All this only means that you recognize the natural order and fitness of things which teach that, in any particular department of life, you must seek for results in accordance with causes that naturally belong to work in that department. Apply this principle now to our subject.

A man desires a profitable crop of wheat in a certain field. The laws and forces of agriculture are divine laws, and they never change. Let the man, then, see to it that his field is one naturally adapted to the raising of wheat. Then let him look after tillage and dressing, and the quality of his seed, and its proper sowing. In all ordinary seasons he can thus make his harvest sure. But prayer, a spiritual force, has no relation to sunshine or rain or frost, or any of the laws of agriculture. And to expect that he can neglect his proper work, and then guard against or retrieve his failure by prayer, is simple impiety. It is expecting that God will put a premium on laziness and ignorance and neglect, — crops which the divine Husbandman does not care to grow. Through the physical laws of my body, God speaks to me with a voice as sacred and commanding as that said to have been heard on Sinai; and he says, Obey these laws, and you shall have and keep your health. Ignorantly or wilfully I break them, and am sick. God still says to me, These my laws point out to you the divine way back to health; but, still neglecting these, I expect to circumvent God's own methods, and get back to health again by a superstitious talisman, a saint's relics, or some ignorant enthusiast's prayers. It is just as presumptuous, impious, and foolish, as it would be for me to jump off Bunker Hill Monument, and then pray to God to suspend gravitation and make my fall easy; or as if I should touch a match to a keg of powder, and ask God to keep it from exploding. The laws of health are just as fixed and certain, and just as divine, as those of powder or gravitation. Moody teaches that a minister has no right to look after his worldly affairs; and says that when he wants a barrel of flour, he asks God for it, and it comes. This is all very well so long as Mr. Farwell of Chicago believes Mr. Moody is doing a good work, and stands ready to back him up with both money and flour; but it is nonsense to suppose that prayer would bring the flour if no one had any confidence in the utility of his work. All over the world, and in all time, men and women and children have hungered and cried for bread, and starved with a prayer as their very last breath. Is Moody a special pet, that only his prayers are answered? A man wants a factory on the bank of some running stream. Can he build it with prayer, or out of logical syllogisms, or emotions of the heart? He can build it only in the use of natural forces, and in accordance with natural laws. In the building of the dam, the raising the walls, the construction of the machinery, the adaptation of wheels to the water-power, every step must be a knowledge of natural laws, and a rigid obedience to them; and just in accordance with the knowledge and the obedience will be his success. He combines and adapts laws and forces, and so produces results that otherwise would never have come to pass. Man is no idle spectator of God's working in nature; or, if he were, there would be no civilization. So man does "interfere" with and materially modify the natural order; but so far as he succeeds he interferes with law lawfully. He combines, adjusts, and adapts, and so accomplishes his results. All the forces that are represented in a train of cars, or the Atlantic cable, are natural, divine forces: but nature alone would never have made either. And yet man does not contravene the laws in making them: he combines simple forces for the production of a complex result.

But all these things come through obedience to God in the special department where the result is reached. Mr. Frothingham has been cried out against for saying that the popular notion of prayer is immoral; but a little thought will teach you that he is right. To disregard God's methods in one department of life, and expect to escape the consequences by resort to the methods of another department, is nothing more nor less than to expect that God is going to take one hand to deliver you out of the other. Men say practically, "O God, I will not obey your conditions of health; but I expect that when I pray you will make me well. I will not pay any heed to your law of gravitation; but you must keep me from falling and being injured. I will not regard your laws of steam; but I hope you will make my engine work just as well as if I did. I will not study to know the laws of gas; but I trust, for the sake of my prayers, you will make my house as light as my neighbor's. I will not pay any attention to your laws in the strength of materials; but I pray that my store may stand as strong and as safe as any on the street. I will not keep your laws of morality and character; but I will become as good as anybody else by praying that you will suspend the general rules in my particular case." That is just what the common idea of prayer means. It puts a premium on laziness and ignorance and incapacity and wilfulness. If it is true, there is no need of knowledge, of labor, of training, of skill, of foresight, of care. In the truest and deepest sense of the word, it is immoral. Carried out logically it would make civilization impossible. What is the use of merchandise if God brings flour to the door of every man who asks for it? Mr. Moody ought to go a step further, and save his wife the housekeeping trouble by having the flour ready made into bread. On his theory, a breath could do it. What is the use of skilled physicians when prayer alone can heal the sick? What is the use of trained captains and drilled sailors, much more of insurance companies, if prayer will always insure a safe voyage at sea?

These natural conditions and laws are the present, active, working God. He who knows and obeys the conditions becomes master of the divine omnipotence. The whole force of divinity helps him.

Must we not, then, pray? If there be spiritual life in you, you cannot help praying, any more than a rose can help exhaling its fragrance. The child does not sit dumb in the presence of father and mother, because it knows the love and care of his parents do not depend on regular asking for them. The child-heart seeks rest and love in the parent-heart, and naturally pours out its thoughts, hopes, fears, and wishes, into the sympathetic ears; but, if the child be a wise one, he does not expect to make his wishes prevail against higher and better wishes.

And yet the prayer may be a very vital thing in the matter of our character and relation to God. What does

compliance with divine conditions do in nature? Simply
this: It does not change any single law or force; it only
sets us in new relations to them. Gravitation will hold
me firmly on my feet, or it will fling me down an abyss,
according to the relation in which I stand to it. When I
obey them, laws help me: when I disobey, they hurt me.
Thus prayer may set me in new and higher relations to
God, so as utterly to change, and grandly to elevate, my
character. When by spiritual, aspiring prayer, I reach out
after God, I comply with the conditions of spiritual health
and strength. If I open my shutters toward the east, the
morning sun will shine in. It will shine any way, but will
do me no good unless I obey the conditions of its shining
on me. So, if I open the windows of my soul toward
God, the light of his divine truth and life will shine in.
In this spiritual realm it is knowledge and obedience to
divine laws and conditions, precisely the same as in the
material. It is one God and one order in both. Study
and work, then, are material prayer; and prayer is spiritual study and work. "I will, therefore, that all men pray
everywhere."

> "For, what are men better than sheep and goats,
> That nourish a blind life within the brain,
> If, having hands, they lift them not in prayer
> Both for themselves and those that call them friend?
> For so the whole round earth is every way
> Bound by gold chains about the feet of God."

So far, then, as both natural and spiritual are concerned, when we make the request, — "Lord teach us to pray," the answer comes, that the way for us to find God, and get his forces as helpers on our side, is by knowing and obeying the divine will — the laws and conditions — in whatever department we wish the results produced. Thus the highest prayer is, "Not as I will, but as thou wilt." God's will is the only power; and it works out our purposes when we have obeyed the conditions, so that the divine forces flow in the channels that we have intelligently dug out for them.

> "And yet the spirit in my heart
>   Says, Wherefore should I pray
> That thou wouldst seek me with thy love,
>   Since thou dost seek alway;
>
> "And dost not even wait until
>   I urge my step to thee;
> But in the darkness of my life
>   Art coming still to me?
>
> "I pray not, then, because I would:
>   I pray because I must;
> There is no meaning in my prayer
>   But thankfulness and trust.
>
> "I would not have thee otherwise
>   Than what thou ever art:
> Be still thyself, and then I know
>   We cannot live apart."

IX.

BIBLES, AND THE BIBLE.

The clothing, the buildings, the institutions, the arts, the commerce, the rites, the ceremonies, the books, the general habits and customs of a people are the natural outcome and expression of that people's life. So all these things change as the people change. They rise, they progress, they decay, as the people rise, progress, or decay. None of these things, then, can be permanent in any special form, unless humanity could stagnate, and come to a permanent standstill. And since we believe that man began at the animal level, and is rising as the ages advance, we must also believe that none of the outer manifestations of the life of man has yet reached its completest manifestation. The art, the architecture, the literature, the statesmanship, of barbarous man, are all barbaric. As he develops, they develop and take on new forms. And unless we believe that humanity has gotten its growth, and will never be any wiser and stronger and better, we must expect that all our outer life — that we comprehend under the word "civilization" — will in the future

> "Suffer a sea-change
> Into something rich and strange," —

becoming richer and finer than we now know; and, though strange to our present thought, yet familiar to that which shall be. But all this change will be in the order of growth; new and strange only as manhood is new and strange to a child.

This law of change and growth, which is true in all other things, holds also with equal force in matters of religion. The religious rites, institutions, and books of a people, are and must be the natural expression of that people's religious thought and grade of civilization. In rude and savage nations, where the art is grotesque, and the whole life is on the barbaric level, the religious thought and form correspond. The images and pictures of the gods are rude, the cultus is coarse, sometimes obscene, always rude and gross; and the hopes and fears find expression in such rough and material shapes as are fitted to impress rough and material people. As civilization develops, these things grow, — a few thoughtful men always ahead of the crowd, — until the cultus becomes stately and grand, the idols masterpieces of art, and the Scriptures the highest religious aspirations and inspirations of the time.

These things, which we know to be historic truth, are precisely what evolution demands; for it teaches that the whole outer life of man must be but the natural unfolding of that which is within him. The religious writings of the nations, that they have come to look upon as bibles, — divine revelations, — are no exception to this rule. They have been, at the time, the natural outgrowth of the people's religious life. That they have been produced only rarely, and at special epochs of upheaval or change, no more militates against their natural development than the fact that the century-plant blooms only once in a hundred years takes it out of the category of flowers. To say that, if bibles are natural developments of humanity, they ought to be more common and in perpetual process of manufacture, is no more conclusive than as if you should say, since Shakspere is not miraculous, we ought to expect to find a Shakspere in every country village.

The rudest religions have no bibles, just as the rudest peoples have no literature. Every race that has become so cultivated as to have a literature has also had its sacred literature, or bible, — the one the natural expression of its religious life, as the other is the natural expression of its intellectual. So the Chinese have the writings of Confucius; the Hindoos have the Vedas and Brahmanic scriptures; the Buddhists, the works of their master; the Persians, the Zendavesta; the classic nations, their hymns and oracles; the Norsemen, their Eddas; the Hebrews, their laws, prophets, and psalms; the Christians, their Gospels and Epistles; the followers of Joe Smith, the Book of Mormon; the New Church, the writings of Swedenborg; and the Spiritualists, the books of Andrew Jackson Davis, and a host of others. One thing, if you will notice it, is common to all. Noble or ignoble, wise or unwise, inspired or uninspired, all these various scriptures have on them the birthmark of their nationality and their time. They are all the natural and necessary outgrowth of the religious thought and life, hopes and fears and aspirations, of their age. They partake of the ignorance and limitations and prejudices of their time. Not one of them betrays any veritable knowledge higher than the high-water mark of their epoch. This means simply that they were growths of the earth, and not exotics transplanted from heaven.

One other thing I wish you to notice, that is true of them all: not one of them came to men with such credentials of revelation as to make them take their place, from the outset, as the "word of God." The sacredness with which they are regarded is a matter of growth. The reverence that surrounds them is chiefly the halo of antiquity. Paul was personally despised, and his letters cast out and abused, by the majority of his fellow-Christians. Only after ages have passed away have books taken on sacredness. As the earthly authorship grows dim and distant, the heavenly claim is brought forward and emphasized. It seems to be true, that the less people know about a religious claim, the more they will believe in it.

And another habit of humanity needs emphasizing. When they once come to believe in the sacredness of a thing, they will not give it up even for something that is manifestly better — as though any stronger proof could be given for the divine authorship of a thing than its truth, beauty, and utility. Take one striking illustration of this: In the earliest times, the use of metals was unknown. The priest then used a stone knife in sacrifice, for the very good reason that a stone knife was all and the best he had. Through use, the stone knife became sacred. So, when metal knives were at last invented, the people dare not use them; and they kept to the worse, because it was old and sacred. How much wiser or more sensibly religious are we? Because the Bible sanctioned slavery, thousands defended and fought for it. Though manifestly inhuman, they thought, because it was in the Bible, it must be divine. And to-day the question of Sunday and our centennial cannot be argued on the ground of human fitness and utility, because of religious prejudice. So all causes must be brought, not to the test of reason, of sense, of experience, of utility, but must come to the Book. And yet the writers of the Book wrote in other times and with other peoples, and with less of knowledge and experience of humanity than thousands possess to-day. What Buddha and Zoroaster and Confucius and Moses and Paul thought and said, are of binding force to-day — *if they are true;* but their mistakes and limitations should not bind us simply because they are in bibles, and are called canonical.

All this will be admitted by everybody, and in all nations, concerning every other bible but their own. The practical question, then, for us to settle, is, as to whether our bible is an exception. All religionists regard their own bible as a divine revelation; and they reject all others. They know theirs is true; and they know all others are not. What, then, is ours?

The common claim of the Church that the Bible is, and ought to be, the court of final appeal, and the end of human reason, rests on the other claim that the Bible is inspired, and so infallible. The foundation of this latter claim, then, is the thing for us to examine.

As specimens of the ordinary argument, and of the use of texts as proof, we will just glance at the two pillar passages that are supposed to support the common belief. In 2 Pet. i. 21, it reads: "For the prophecy came not in old time by the will of man; but holy men of God spake as they were moved by the Holy Ghost." Three considerations take all value from this passage as proof. First, the words "prophecy" and "holy men" are indefinite: they do not cover special men or books with any such certainty as makes us know just what is meant. Second, it is more than questionable as to whether or not Peter is the author of the second letter that bears his name. Third, even if it were settled that Peter wrote it, still it remains to be proved that it is any thing more than fallible Peter's fallible opinion. The other passage is in 2 Tim. iii. 16: "All scripture is given by inspiration of God, and is profitable for doctrine, for reproof, for correction, for instruction in righteousness." This passage, being written long before the New Testament was brought together, of course has no reference to any thing but the older writings. But, more than this, the present translation is incorrect. Bishop Ellicott, a leading Church authority in England, makes it read, "Every scripture, inspired by God, is profitable," &c. You will notice that this leaves wholly unsettled the central question, as to what scripture is inspired by God. Our judgment, then, must be made up from other sources than any special texts.

There is one vital distinction that is frequently lost sight of in discussions of this subject. I refer to the distinction between inspiration and infallibility. Proving that a book is infallibly true does not prove that it is inspired, in the ordinary sense of that term, as used among the churches; and, on the other hand, proving that a book is inspired, does not necessarily establish its freedom from error: but as the two things are so often confounded, or claimed together, any clear view of the theme must notice them both. How stands the case, then, in regard to biblical infallibility?

To begin with, it is a fact worthy of earnest attention, that the Bible nowhere makes any claim to be infallible. We have already seen how little the strongest passages of the kind to be found are able to bear any such strain of interpretation: and, as a fair specimen of the style and real pretensions of the writers, notice the opening words of Luke's Gospel. Here he says, that, as others were writing down their accounts of the life and sayings of the Lord, "It seemed good to me also, having traced down every thing from the first, to write unto thee in order, most excellent Theophilus;" and one "orthodox" commentator remarks upon this, that "inspiration did not render it unnecessary to use every available source of information." Luke merely says that, having had means of information, it seemed good to him to write his Gospel.

And,—what is not true,—even though some one writer should claim to be infallible, and should make good that claim, his infallibility would be no guaranty of the infallibility of any one else; for there is no reason better than the bookbinder's, or simple convenience, why all the books of the Bible should be together in the same covers. What one says, therefore, does not necessarily hold true of, or represent, any of the others.

What are the real facts about the Bible as a book? The different books were composed by different authors, of different nationalities, and at periods so widely apart that the time of their writing stretches across a space of fifteen hundred years. Of some of them, nobody knows when, where, or by whom, they were composed, or how they have come to be in their present places or shapes. Of others, the authorship, though not quite so obscure, is still in dispute; and this not of certain unimportant ones, but of the very central books of the New Testament. The authorship of the Gospel of John is not yet determined; neither is it known that we have either of the Gospels in its original form.

And then, even though we knew who wrote them all,

and where and when, we have no sort of guaranty that they have been handed down with such textual accuracy as to justify any bishop or church in pronouncing judgment on the basis of its present verbal utterances. We have only modern copies of books from two to three thousand years old. They floated about in the hands of persons or organizations for generations before they were collected as we have them now. In copying they were purposely or blunderingly altered — sometimes, we know, to the extent of whole paragraphs: how much more, we do not know. In no case have we an original manuscript. The most ancient one we possess was written at a time farther from Christ than we now are from Shakspere. And when you think of the disputes of commentators over the text of the dramatist, — and this in an age of printing, — and remember that even his personality has been called in question, you can judge something of what probability there is that we can be so sure of the mere words of the Bible as to warrant us in using them as instruments of hatred and warfare, and even of present and future damnation of our brethren.

And even if we knew all about the separate books, and were sure we had accurate copies, we do not know with certainty as to what is bible, and what is not. I mean by this that the sacred canon has never yet been definitely settled, even by the Church itself. The Council of Trent, and all the Roman Church until now, not only declare the whole Apocrypha canonical, but anathematize all who dissent. The whole Protestant world rejects it.

And the best scholarship of the Church is still unsettled about Hebrews, James, Jude, the second epistle of Peter, the second and third of John, and the Revelation.

If there be infallible books, of which to make an infallible bible, and if these be infallibly preserved and transmitted to us, we are still undecided and in trouble, unless we have also an infallible catalogue to tell us which they are. If there are two or three guide-posts, and one is infallibly correct, and the others not, it matters little to us, unless some one is able to tell us which is right. And then, if *words* be so important, how comes it that the New Testament writers quote the Old loosely and incorrectly? In one place, the Septuagint is followed where its translation from the original Hebrew is blunderingly wrong, and even reverses the sense; and not only are these things so, but there are in the Bible palpable errors and inconsistencies and contradictions that no one would think of trying to cover up, were it not for the pressing necessities of special pleading. Not only is it confessedly impossible to reconcile Genesis — I do not say Moses, because it is extremely doubtful as to whether Moses wrote the Pentateuch — with the words of God's great world-book that we know he wrote, whether he wrote any thing else or not; but even the New Testament writers do not always agree, and, in some cases, were most certainly mistaken. I have not time to dwell separately on the contradictions of the evangelists concerning the birth of Jesus, the length of his ministry, the date of the supper and the crucifixion, and the incidents connected with his resurrection. In regard to these, I content myself with the general statement that they never have been successfully reconciled. Let me, however, emphasize the point by two or three examples.

In the first place, I refer you to the genealogy of Jesus as contained in Matthew and Luke, as compared with each other, and with the Old Testament records. The table of Matthew is *not correct:* it is not the genealogy of Jesus, in any sense, unless he is the natural son of Joseph; and it is wholly irreconcilable with the table of Luke or with that of the Chronicles. It has a suspicious look to find that the genealogical table is divided into three perfectly equal parts; and it is found wholly indefensible when we notice that this even division is obtained by changes and omissions. The matter is confirmed by the discovery that, while Matthew gives but twenty-six generations between David and Joseph, Luke gives forty-one; and then, both Matthew's and Luke's tables give the genealogy of Joseph, which is nothing at all to the point, unless Jesus be the son of Joseph.

In the next place, the evangelists quote the Old Testament inaccurately not only, but they quote as prophecies of Christ words that have no reference to him whatever. In one place, a passage from Zachariah is attributed to Jeremiah; and, in another, a passage is professedly quoted which has no existence in the Old Testament, nor anywhere else.

Once more, notice the long address attributed to Jesus concerning the destruction of Jerusalem, and the "end of the world." False Messiahs and great wars were to precede the downfall of the city. We know from history that neither of these things occurred. Christ was to come again in the clouds of heaven, immediately after its overthrow, and before that generation passed away. He did not so come.

Paul was mistaken on this same point. He teaches distinctly, in 1 Thess. iv., that Christ was to come and raise the dead, and set up his Messianic kingdom, before "we who are alive" have passed away.

And then, this doctrine that the Bible was infallible, and not to be touched, criticised, and judged of, by individual students, was not prominently held out till the sixteenth century. Luther needed an infallible book to set up in opposition to infallible Rome, and so claimed he had one; and, while he affirmed the "right of private judgment" as against the pope, he did not hesitate to deny it, even to the point of persecution by the civil power, as against Luther.

Again: Peter and Paul were at swords' points on the question of receiving the Gentiles into the Christian Church. Are we to assume that they immediately became infallible on taking a *pen* in their hands, while they were mistaken about the gravest matters when *talking* and *acting*, and while they were even capable, when hard pushed, of dissimulation or falsehood? And, once more, Jude was not so infallible but that he could quote from the apocryphal book of Enoch, written but a little while before his own time, and all the while suppose himself repeating the words of "the seventh from Adam," who was translated.

"But if the Bible is not infallible," say some, "how are we to get along for a guide? We must have an infallible guide."

I am perfectly aware that this kind of reasoning is much more common than it is logical or honest. Such arguments — and books and pulpits are full of it — are mere appeals to passion and prejudice, that would not be needed were not facts sadly wanting. Never forget one thing: the consequences of a fact can never invalidate it; and that you want a thing is not quite sufficient ground for believing it true. Suppose you thought yourself worth fifty thousand dollars, but wake up to find you have not a cent. "It can't be! I must have money! I hate the man who brought the news!" These and like expressions would hardly add much to your bank account. It is well, it is honest, and it will pay in the end, though temporarily painful, to find out and face the exact truth.

But what am I saying about its being *painful* to face such facts as these? In a certain sense, it is painful to give up any old and cherished belief. It is not pleasant to find that the rock you had anchored to is only an ice-cake, and melting at that. It is such a comfort to feel that you have an unchangeable standard to refer to.

I have two remarks to make about this; and, first, all this talk of a fixed standard in the Bible, outside of the last decree of the Church of Rome, is pure illusion. There is hardly a doctrinal passage in the Bible that "orthodox" Christendom is agreed about. It is hardly satisfactory to a person seeking exact truth, to have half a dozen doctors of divinity each giving the only correct interpretation of a text, and all anathematizing the other five. Looked at through the medium of denominational commentators, the various doctrines assume as many shapes as the cloud in Hamlet: —

Hamlet. Do you see yonder cloud that's almost in shape of a camel?
Polonius. By the mass, and 'tis like a camel indeed.
Ham. Methinks it is like a weasel.
Pol. It is backed like a weasel.
Ham. Or like a whale.
Pol. Very like a whale.

All this talk of an infallible standard is meaningless and delusive, unless you have, as Romanists claim, an infallible interpreter.

And the second remark is this: So far from its being painful to give up the doctrine of biblical infallibility, I hesitate not to say that, just in so far as one has the spirit and love of Christ, and at the same time comprehends intelligently the points at issue, he will rejoice to be able to give it up; and what little comfort there may be in thinking he has an infallible guide, he will gladly sacrifice to the larger and better results. So far from there being any justification for "orthodox" writers and preachers, when they say that those who question the infallibility of the Bible are undermining the hope of man, it is true, on the other hand, that bald and blank materialism gives a nobler, juster, and more blessed picture of human life and destiny, than does "orthodoxy." He who has in his heart something of the love and devotion and self-sacrifice of Christ would rather take the boon of his brief life beneath the stars, and then at its close go down into darkness and silence and nonentity, than to grasp eagerly after the gift of immortality, if, along with that, he must also take the appalling fact of the unending sorrow and wail of one single human soul. I do not say that our feelings or preferences touch the fact of the matter in any way whatever: I am talking of the question as to which side has the right to charge the other with teaching horrible and hopeless doctrine. I would rather die like a dog, and see no to-morrow, than selfishly take a heaven while one single soul is in endless torment, even on the outermost verge of creation; and I do not envy that man his Christian charity, who would not surrender his immortality for the sake of the abolition of hell.

Am I not right, then, when I say that the "orthodox" hope of heaven, alongside an everlasting hell, is more horrible than the materialist's hope, that, though it do not contain any heaven, is also destitute of a hell?

A belief in the infallibility of the Bible is one of the greatest evils of our Christianity and civilization. It perverts and makes childish, partial, despotic, and horrible, the character of God. It misrepresents and maligns the nature and destiny of man. It warps moral perceptions, clogs progress, and hinders civilization. It makes the churches, instead of reformers and evil-slayers, sectarian and jealous clans in perpetual feuds. It turns them into system-makers and riddle-guessers and Chinese-puzzle-arrangers. In short, not one single valuable thing will be lost when the fiction of infallibility is surrendered; and many and grievous evils will be thrown off.

So much for infallibility. I have given time to it because of its importance, and because of the ignorance and prejudice current concerning it. I will despatch the second part of my subject as briefly as I can, though I do not wish to slight it on account of haste. Turn now, then, to inspiration.

This, as I have already said, is quite a different thing from infallibility. The Bible writers do claim for themselves inspiration, — that they speak and act under a divine guidance and influence.

Before proceeding to investigate and define this claim, let us glance at the different theories of inspiration that have obtained. By so doing, we shall find that we have already incidentally replied to some of them; and so we shall clear our way, and see what path we are to follow.

And, first, the prevalent theory, from the time of the Reformation until forty or fifty years ago, was the one called the verbal. This taught that every word of the Bible was the direct communication of God to the mind of man, and this as much as though God held the hand and directed the pen of the writer. Old Dr. Owen, a prominent English Puritan divine and commentator, held that even the points and accents of the original Hebrew — which we now know were invented for convenience' sake, long after the time of Christ — were divinely inspired. And as these masoretic points, as they were called, stood for vowels in the Hebrew tongue, and as a different point or vowel might make a different word, he rightly held that, unless they were inspired, he had no certainty as to the meaning. Owen's successors have been compelled to surrender his premises; though they still illogically cling to his conclusions, so far as infallibility is concerned. But as verbal inspiration is not now claimed by any, so far as I know, it is not needful to give it any further attention.

The theory that followed that, and is the "orthodox" one of the present day, is that which is called plenary inspiration. It gets its name from the Latin *plenus*, meaning full, complete. It teaches that the Bible writers were so inspired as to put them fully, completely, in possession of truth, and enable them to teach that which was wholly free from error.

This theory has already been overthrown by the same arguments and facts that overthrew the doctrine of infallibility; for of course, if there are any mistakes or false teaching in the Bible, that proves that the writers were not possessors of plenary inspiration.

There is one more theory left, which may perhaps, with sufficient propriety, go under the name of the Liberal. This theory teaches that every soul — being made in the image of God, and so capable of communion with him — is open to the divine influx, and does in its measure receive it, just as every gulf and bay and inlet and shallow and sinuosity, on the ocean-shore, is open to the inflow of the tides. But human capacity for the reception of this inspiration is conditioned and gauged by constitution and character: so that as, on the seashore, the openings for the sea range all the way from gulfs to shallows, so the capabilities of human souls mark gradations that reach all the way from the evanescent impulse and uplifting of a common man in common daily life, up to the magnificent spiritual insight of Paul. And it holds out the hope that, by faithfulness to what we have, we may grow to an ever-enlarging capacity, on the principle that "to him that hath shall be given," so that by and by we may walk under a whole heaven full of light, and read the law so perfectly written on our hearts that we shall be a law unto ourselves. Perhaps I need hardly say that this is my own belief.

Before proceeding to speak of its sufficiency or value, let me indicate to you what seems to me its reasonable, scientific foundation. However great the havoc that science has made with the forms and supports of theology, it has given utterance, through its most distinguished mouthpiece, Herbert Spencer, to what it confesses must be an eternal basis for science not only, but also for religion. It shows us how all things run back and down into an underlying, changeless, and yet inexplicable force and life. This life or force science may call one thing or another, as it will. A name cannot annihilate it; and we will call it God. This life and power of God is the only possible explanation of any thing. How does the grass grow? Science answers, "Nature, force;" religion answers, "God." How came the stars in heaven, and how comes their light to us? Science says, "Nature, force," once more; and religion again replies, "God." How came man to be, and whence his powers of thought and love? Once more science repeats, "Nature, force;" and once more religion says, "God."

This unseen and incomprehensible God, which is a Bible doctrine as well as a teaching of Spencer, is the life of all that lives, and the motion of all that moves. Every good and holy thought, every noble deed, every high endeavor, is by and through so much of God as works through humanity; for without him we can do nothing. He dwells in us, both to will and to do. "In him we live and move and have our being." This is nothing more nor less than the doctrine of the immanence of God in the universe and in man, — all things by God, and through God, and for God.

Inspiration, then, is natural to the human soul, and its degree is determined by character and capacity; and it is not confined to the teachings of formal religious truth. Even the Old Testament teaches that certain men were inspired of God to work in linen and brass and cedar and gold. Why not, then, Shakspere and Michael Angelo, and Socrates and Epictetus, make good a claim to think and work by the inspiration of God?

Let us see. All truth, of whatever kind or degree, is from God. All light is from the sun. Whether it shine from moon or planet; whether it be reflected by brook or mirror; whether it stray, a broken beam, into some prison-cell; whether it flare in the gaslight, or glow in the coal of our evening grate, — all light is, first or last, just so much sunlight. So, whether a truth be in Bible or science, in Christianity or Paganism, on the banks of the Jordan or the Ganges, wherever found, and by whatever path it come, being true, it must be from God.

The Bible, then, though not in any exclusive sense the word of God, yet does contain the word of God; and, in so far as it is true, so does any and every other book; and every book, so far as true, is sacred, as being a reflection of the divine.

Do I, then, put the Bible on the same level with all other books? By no manner of means. Here is a primer, and here is Sir Walter Scott; both are books. Does saying that degrade Sir Walter to the level of the primer? Here is Tupper, and here is Shakspere. Because you call them both poets, does that bring Shakspere down to Tupper? Does calling a third-rate painter an artist make him equal to Raphael? I would still put the Bible apart by itself, as, in a certain grand and real sense, what the ages have agreed to name it — the Book. I would do this, because, after all that has been or can be said, it still, by the degree of its inspiration, and by the quantity and quality of truth it contains, outranks the literature and religion of the world.

What I have said has not been against, and does not invalidate the work of, the Bible. It is not against any claim that the Bible makes. It is only against false and unfounded claims made on its behalf, — claims, the nature of and failure to prove which, are constant hinderances to a perception of its real worth, and so injurious to the simple, magnificent, and divine truth of Christianity.

The central teaching of the Christianity of Jesus, contained in the Bible, is the true theory of humanity, and the inspired truth of God. Much that has passed under its name is superstition, Pharisaism, and Paganism. The truth of the latter statement may be easily proved by study and comparison. The truth of the first may be verified by reason, and demonstrated by trial. Jesus' conception of God is the grandest that the mind of man can yet conceive; and his theory of human life is the outline of the best we can hope for. The underlying principles of Christianity, and the underlying principles of humanity, as it moves on toward perfection, are perfectly identical. This alone is demonstrative of its truth; and that it is true is the highest of all conceivable authority, and is absolute proof that it came from, or was inspired by, God. Any amount of evidence or human testimony could not be so authoritative.

So far, then, as the central truths of Christianity are concerned, here is certainty beyond any thing that any theory of "verbal" or "plenary" inspiration could give.

As for that further question, so often asked, How am I to know whether a particular teaching is, or is not, true, if all the Bible is not to be received? I reply, By the

general consent of the highest moral sense of the world."
How do men test works of art, and assign them their
rank? By the general consent of artists, cultured men,
and those of the highest taste and sense of beauty.
There is no standard that one can carry in his pocket,
that is able to guide a boor in the selection of a fine
painting. He who has taste knows at a glance. So no
amount of Bibles can help unspiritual and selfish men to
discern spiritual truth. Can a book in the pocket or on
the parlor table of a miser help him discern the star-like
glory of "It is more blessed to give than to receive"?
How many inspired volumes would it take to help a rake
see the truth of "Blessed are the pure in heart"? The
purest, best, and noblest men of the world are the highest
standards of moral and spiritual truth, just as the greatest
statesmen and jurists are standards in their specialties,
or as the greatest poets and artists are standards in
theirs. Shakspere is a criterion by which to judge the
world's dramas; and Titian and Angelo are the last appeal of artists. They are all we have; and they are
enough. They bound the horizon of human vision in
the direction of the beautiful. And so Jesus and John
and Paul are moral and spiritual seers. They overtop
the mass of the world, and gaze afar. They bound
the horizon of earth in the direction of character and
God.

Then this inspiration is not something once given for a
while, and then forever after withdrawn. It is too great

for any book; it breaks out of the Bible, and becomes a living stream, to follow us in all our lives, and humanity through all time, as the brook that burst out of the rock, under the rod of Moses, is said to have followed and quenched the thirst of Israel in the wilderness. It says, "Lo, I am with you alway, even unto the end of the world." God is not shut up in covers. "The word of God is not bound" by the bookbinder, any more than in any other way. Neither is it dependent on the Bible, like a physician on his medicine-chest, so that if that be gone he can do no healing work. All heathendom is not "without God;" for in "all nations those that fear God are accepted of him." He is a living, loving, guiding spirit. He gives light to the eyes, and strength to the heart, of the nations. He has more truth than is in the Bible; and the process of the ages is but the unrolling of his divinely written scroll. What matter, then, though we do not certainly know each step we are taking? Are the children of a ship-captain less safe because they do not understand the log-book, the quadrant, the path of the vessel through the waves? A wise head and a loving heart are in the cabin, and a strong and wakeful hand is on the wheel. The captain knows where he is going; and he knows his route; and the smallest, weakest, and most ignorant child shall go sailing up the harbor, and when the anchor is dropped, and the boat lowered, shall set foot on the wave-washed, sandy beach of the everlasting shore, just as surely and safely as the captain himself.

Have faith, then, not in churches, nor creeds, nor councils, nor books: "have faith in God;" for

"I doubt not through the ages one increasing purpose runs.

. . . .

Not in vain the distance beacons. Forward, forward let us range;
Let the great world spin forever down the ringing grooves of change.
Through the shadow of the globe we sweep into the younger day."

I have a great deal of doubt of men,— their thoughts, their creeds, and their systems; but with all my heart and soul I believe in God and the future. He has inspired and led in all the past; he inspires and leads to-day; he will inspire and lead to-morrow; for "he is the same yesterday, to-day, and forever."

X.

THE DOCTRINE OF ATONEMENT.

THE atonement is the central doctrine of Christianity, and of right and necessity is it so; for, if the fundamental principles of the popular system be once admitted, then the necessity of this doctrine of atonement logically follows. Without any *the*, it is the central doctrine of all religion. It always has been, and it always will be. To understand the truth of this statement, let us ask what atonement means. Atonement, of course, can have relation to the situation in which two human beings stand to each other. If there has been estrangement between friends on account of an injury rendered one by the other, or on account of an innocent mutual misunderstanding on the part of both, *atonement* means bringing them together again, reconciling between them, clearing away the misunderstanding, and bringing them once more into the position of their old sympathy and friendship. Of course, from the religious standpoint, atonement must always have relation to the position which humanity occupies in regard to God. What does atonement mean here, then, and what shape must it necessarily assume? It will be determined always, in the character of its definition, by the conception which any special age in the history of the world shall have of God, of humanity, of the actual relation in which they stand to each other, and of the ideal relation in which they ought to stand. If you will think of it a moment, you will see that you must understand these four points before you can get at the conception of the doctrine of atonement, as held at any particular time in the history of the world: What is God? what is man? what relation do they actually sustain? what relation ought they to sustain? And then *atonement* comes in as the means by which the actual is changed into the ideal, — by which the fact becomes what ought to be the fact.

If I should prove my statement, that atonement is the central doctrine of all religion, the bringing of God and man at one, it would involve the necessity on my part of giving you the universal history of the religious thought and the religious life of the world. Of course, within my present limits, I can only point out two or three of the more important stages in the development of this thought, along with the applications that will naturally follow from them. Following, then, the line up the pathway of humanity by which the progress has been attained of which we are to-day the representatives, and coming down from the most ancient times to the present, you will recognize three distinct and separate stages in the thought of man concerning this great doctrine of atonement. Go back to the earliest times. Atonement then possessed nothing whatever of any moral element, or what we should now call "religious." What was the conception of God, the conception of humanity, and their relations by which the form of the doctrine was determined, in those earlier ages? The gods themselves were not moral beings; they did not represent any moral idea; they did not stand for the human conception of righteousness. They were simply the personified forces of the world; and the relations which men sustained to those gods were not moral relations. Men never bowed down before Jupiter, or before Odin, or before Brahma, with the consciousness of sin such as men speak of to-day, and with the desire to be made purer in heart, and reconciled to this god, who to them was the ideal of righteousness and truth; for, as I have said, those old nature deities were not the representatives of righteousness and truth. They were simply forces, powers, having a certain influence over men, being able to help or to damage them, and which they stood in fear of, or whose favor they wished to gain. So that atonement, as I said, in this first stage of the world's religion, possessed no moral element; it was not a reconciliation between human unrighteousness and divine righteousness. What was it, then? Suppose there came some dire calamity upon a person, a family, or a nation, so that they conceived that some one of the gods was angry: what did they do? Why, they endeavored to find out, by the best means they had at their disposal, what it was that they had done, and why it was that they had incurred the divine displeasure; and atonement, to their minds, was simply an endeavor to placate the wrath of offended deity. It meant building him an altar, erecting a temple, bringing to him offerings, burning sacrifices, pouring out oblations, establishing rituals, — doing something by which he should be appeased. Or, if one of the gods was indifferent, and they wished to gain his favor, it was simply bringing to him some offering that they supposed he would desire, that would enlist his interest on their behalf.

Then, as I have said, the first form that the doctrine of atonement assumed (this attempt to bring humanity and God together) was not a moral form, and had nothing whatever to do with righteousness or with morals; but out of this conception of the Divine, and of man's relationship to him, grew one of the most important elements of all religions, an element that has not been outgrown, an element that largely forms and gives color to the popular Christianity even of to-day. I refer to the element of sacrifice, not as the giving of something for the attainment of something else, but sacrifice as a religious ceremony, — burnt offerings, the bringing of the fruits of the field, the firstlings of the flock, as offerings to God. This idea of sacrifice was once universal; and it grew out of a conception of God as a cruel being, who delighted in suffering, who rejoiced in the smell of the burning fat of the victims that were consumed upon the altar; a God who was ready to take from you your sacredest and your dearest, as a punishment for something that you had done by which you had given him offence. I say this world-wide doctrine of sacrifice, this doctrine, false in its form (for the doctrine of sacrifice itself is universal and eternal), — this false doctrine of sacrifice grew out of this false and partial conception of God, and of his relation to humanity.

The second stage in this development of the doctrine of atonement is represented by the growth, in humanity, of the moral conception of God. It is represented well enough, and perhaps best of all for our purpose, in the later life of the Jewish people; for at first the gods of the Jews, just as well as the gods of other nations, were nature-deities, and did not represent the ideal of righteousness as Jehovah came to represent it in later ages. There grew up, I say, a conception of God as a moral being, as representing the highest ideal of righteousness and of truth; and there grew up, along with this, a conviction of sin on the part of humanity; for, when once their grandest ideal of righteousness was placed on the throne of the universe, then there came into their hearts a conviction that the one thing that God demanded of them was righteousness in heart and in life. But they still attempted to satisfy this moral demand by the old, false form of sacrifice; that is, when a man felt a conviction of sin in his heart, recognized the fact that he had not lived out the divine ideal of righteousness which he felt was demanded by the law of his own conscience, not knowing what else or what better to do, he still brought to the altar of his God an offering, the firstlings of his flock, or the fruits of his vineyard. He still brought some gift, as though he could take some of the things that belonged to God, and, by giving them to him, make up for his own moral deficiencies. This was the weakness, the central weakness, of the old Jewish system of religion; and this began to be recognized after a time by the highest thinkers, and the best religious teachers, of the Jewish nation. You will remember, perhaps, how in the first book of Samuel it is recorded that Saul was commanded by Jehovah to do a certain thing, and did not obey; but afterward, to make up for the deficiency in his own character and his own obedience, he brought large offerings to the altar of God; and then the prophet Samuel rebuked him, saying, "Has the Lord as great delight in burnt offerings and sacrifices as in obeying the voice of the Lord? Behold, to obey is better than sacrifice, and to hearken than the fat of rams;" thus echoing this universal and world-wide consciousness that liberal religion stands for to-day, and stands for, in a certain distinct way, above and beyond all others, — that obedience, righteousness of character and of life, are the things that are acceptable to the Divine. This same idea comes out in the writings of David, in the touching and beautiful psalm where he says, "For thou desirest not sacrifice, else would I give it; thou delightest not in burnt-offering. The sacrifices of God are a broken spirit: a broken and a contrite heart, O God, thou wilt not despise." And then, in the later prophets, particularly in Micah, it is brought out more clearly still, that the one good which God demanded was righteousness, and not the offering of "thousands of rams, or ten thousands of rivers of oil." "To do justly, and to love mercy, and to walk humbly with thy God," is the thing that the prophet demands as the way by which we can bring humanity into accord with the divine life. This old doctrine of sacrifice is tried in the realm of morals, in the effort of humanity to atone for its sins, and to make reconciliation between itself and God, — is tried, and is found wanting.

The third great form of the doctrine of the atonement is that which is represented in popular Christianity. In order that it may be understood, I must give, just as I did in the other case, the fundamental conception of God, of humanity, and of what people supposed human salvation meant. There had grown up in the later Jewish life a belief in immortality, — a belief which was not apparent in their early history and teaching. There had grown up the doctrine of the fall of man, — a doctrine also not apparent in the earliest teaching and belief of the Jews. It was believed that, on account of this fall of man, death had come into the world; that, if man had not sinned and fallen, he would have lived here forever. It was further believed that, on account of this fall, man had come under the wrath of God, into the hands of Satan and his angels: so that after death, unless there was some redemption found for him, he must endure the misery and torment of a never-ending hell. The one thing, then, according to this conception of God, of the nature of man, and of their relations, — the one thing that was needed was some power, not necessarily touching the present life of humanity, but some power that should reach beyond the grave, that should grasp humanity in its fallen condition, rescue it from the clutches of the Evil One, save it from the wrath to come, and, delivering it from its dark destiny of abiding in the midst of torment forever in one place, open for it the gates of the city of life, and suffer it to enter into the bliss and joy of God forever. This was the problem to be solved. How was it solved? There grew up the doctrine that Christ was the atoning sacrifice that had been prefigured through all ages, whom all the world had looked forward to, and who now had wrought the one work by which humanity was to be saved in the future life. That there is no consistent revelation in regard to this Christian doctrine of atonement, that makes it binding on the thought of all honest persons, is apparent from one or two very remarkable facts. In the first place, you may search all through the history of the words and teachings of Jesus himself, and you may search in vain, to find any doctrine whatever of atonement, as it is popularly understood to-day. There is not one single word that teaches or supports it in any accredited utterance of Jesus; so that we are brought face to face with this remarkable fact: if Jesus did come into the world to bring about such results as are claimed, he utterly failed to speak one word in regard to the great central object of his mission, — apparently forgot entirely the burden of the message he had been sent to deliver. And the other remarkable fact is this: that the doctrine of the atonement, as held by Christendom, has itself undergone changes and transformations, and has followed the line of a definite progress from the first until now. It has assumed some fifteen or twenty different shapes, during the history of Christendom. In the first place, Jesus was supposed to be a price paid to the devil for the purchase of humanity. That is, humanity, by sin and the fall, had become the property of Satan, he having a right to own and control it. Jesus was his ancient rival in heaven, as represented in the poetry of "Paradise Lost;" and so there was a bargain made, by which Jesus was to be delivered over into the hands of Satan, that he might work his will on him, in consideration of the salvation of certain numbers of the human race, who were to be redeemed by this purchase, and who were to take the place vacated by him in heaven. But Jesus being divine, so that, as the New Testament says, he could not be "holden of death," when he had descended into hell, broke away from the power of Satan by which he was held, burst the gates of Hades, and escaped, leading a great multitude of followers, who for ages had been chained in the dark abysses of despair: so that Jesus, by his superior wit and his superior power, outwitted Satan, and delivered not only himself from his clutches, but all those who thereafter should believe on him. This was the first form of the doctrine of atonement, which was held in the Church for ages. You will find traces of it all through the New Testament, if you read it in the light of this fact; and we read it in the

Prayer-Book to-day. It is recited in the Episcopal, the Anglican, and the Catholic churches every day, how Jesus actually descended into hell, and how he delivered himself thence.

Then the doctrine of the atonement was changed, and took on the form of Christ being a sacrifice to appease the wrath of an angry God. This was held, also, for ages. Then it took on the form that is prominently held in what is called "New England theology" to-day, — the doctrine that God somehow must sacrifice Christ as a governmental necessity. That is, here was this great moral government of God, on which the life and safety and peace of humanity depended; and it was claimed that he could not forgive humanity, and still maintain the dignity and the power of his moral law intact, unless there were found some sacrifice, some one who would give himself voluntarily as a victim, so that it might be made manifest to the universe that no man could disobey the laws of God with impunity; and so Christ was found, and he gave himself thus. And then the doctrine has taken on the form of substitution, that has been preached by the evangelists very forcibly and very prominently during the last two or three years. Christ became the purchase for the elect, — a certain part of humanity chosen from all eternity. Christ came, and suffered the precise amount which they would have had to suffer, if they had been chained in darkness and torment through all eternity, and thus he purchased their redemption; and they, through the power of God and by faith in him, are redeemed and set free, and permitted to enter on the divine life.

The last form that I care to notice at this time, and which is a recent one, is that which is received and promulgated by the younger clergymen of the country; the one that is popularly held and taught by what is called "liberal orthodoxy;" the one which is intimately associated with the name and fame and life-work of the celebrated Dr. Bushnell, recently deceased; the doctrine that Christ, by his life, by his teachings, by his sufferings, by his death, revealed the righteousness and love of God, and became a power for the conviction of sin, and an impulse toward a life of righteousness on the part of humanity. This is the moral view of the atonement; and this, I believe, contains in it a central kernel of truth, of power, of life, that will abide forever: for Christ, by his life, by his teachings, by his sufferings, by his death, did reveal the majesty and the glory and the love and the righteousness of God; and he did convince men, when they contrasted their life and character with his own, — he did convince men of sin; and he has become a moral impulse to renovate the life of thousands and thousands of men, and will become such an impulse still in the ages of the future.

The doctrine of the atonement, that I believe to be the one that shall be permanent and universal in the future history of the race, does not slough off and leave behind this moral view of the atonement which I have just described. It takes it up into itself, and carries it along as part of the universal and world-wide doctrine: for the moral power of Christ for the renovation of humanity is not expended; and it does not depend at all upon the theological conception of Christ's person, or his nature, or his birth, or his death, or his resurrection. Christ has become part of the moral and religious life of humanity, a part that cannot be eliminated, a part that shall increase, and become mightier and mightier yet, as it takes its proper place in the history of the religious thought and life of the world.

What, then, is the universal doctrine which I believe must bear sway in the future? And here, again, I must start with new definitions. We have rejected the doctrine of the fall of man. It is not something that we have speculatively cast aside for the simple reason that it does not please our thought, or because we have chosen to take up with some new-fangled notion. The doctrine of the fall of man is proved, absolutely proved, to have been untrue; and precisely the reverse of that is established as a matter of fact and demonstration to-day. Humanity has never fallen. Humanity has never been so high, so noble, so pure, so true, as it is to-day. Starting in the dust, it has, through struggle and sorrow and toil and tears, climbed up to its present position of grandeur and of glory, where it catches an outlook of the future that is cheering, inspiriting, and divine. Of course, then, the fall being given up as a part of the doctrine of the religion of to-day and of to-morrow, there is no necessity for any contrivance, on the part of God or man, by which the supposed fall is to be retrieved. The old doctrine of atonement has no place in such a conception of humanity as leaves out the doctrine of the fall; and there is one weak position of what is now called "liberal orthodoxy." I know a great many ministers, and a large number of persons, who still claim to be orthodox, who have totally rejected the doctrine of the fall; and yet, illogically, they retain the doctrine of the atonement: for, if there was no fall, there is no need of atonement, in the theological sense. Then, the salvation that humanity seeks to-day has undergone an entire and important change. It is no longer a salvation simply from future peril. The salvation we need is a salvation that shall take us right here to-day; that shall take us in our homes, that shall take us in our social life, that shall take us in our business life, that shall make us faithful in all departments of our career. This is the salvation that the age is longing for, and reaching out after the attainment of. We care very little to-day, comparatively, as to the chances that await us when we have passed beyond the veil that separates between this life and the next; for we know that if as true, faithful men and women, clean-hearted and clean-handed, we can pass under that veil when the divine hand has lifted it, we may stand unabashed in the presence of our loving Father, who has led us through this life, and has taken us to himself. The salvation we need then is a present salvation.

And one more change has come. It is no longer a salvation of the soul simply: it is a salvation of the *man*.

We recognize no longer any soul that man carries about with him as a piece of property, that he can get insured, that he can provide for the safety of in all contingencies and changes in the future of life. We recognize no division in human nature. Human nature is one; and body, mind, and spirit are only different forms of the manifestation of this one conscious life that we call self. The salvation that is needed, then, is not salvation of the soul: it is salvation of the body; it is salvation of the mind; it is salvation of the spirit; it is something that regards humanity as a unit, and that seeks to save man all over and all through.

And there is another change. There has been growing lately in the thought of the intelligent part of the world, and it is to grow more and more in the future, a doctrine which, in philosophical language, is called the "solidarity of humanity;" that is, the doctrine that not only the individual is a unit, but that humanity collective is one, and that there is no possibility of saving myself, or saving yourself, while we leave the rest of humanity "out in the cold." Humanity is a unit: we must go up or down together. We recognize it sometimes, — wake up to it occasionally here in Boston, or in the other great cities of the world. Now and then there comes the sweep of a pestilence: what does it mean? It comes creeping about the doorways of the great, looking into the windows of the rich, striking down the fair, the beautiful, the educated, the cultivated, from the highest circles of society. What does it mean? It means that the culture and education and wealth of a city have neglected the lowest, the poorest, the outcast, and the mean. It means that somewhere in the city there is collected a centre of ignorance, a centre of crime, a centre of filth, a centre of neglect, that has turned at last into a centre of disease; so that the breath of God's heaven, as it sweeps over the city, has taken this suffering and plague that was neglected and left on one side, or in one spot in the city, and has swept the infection into every house and into every home. It means that the city was one life, and that the head of the city could not with safety neglect the feet; that the upper circles of the city could not with safety to themselves neglect the foundation; that the city life was one; and that there is no possibility of saving the upper while neglecting the lower. This lesson has been taught, if one will look for it, hundreds of times in the history of the world. Take the old Roman civilization, surrounded on every hand by barbarism. By and by the grandeur of that classic life, that classic thought, that classic writing, that classic art, is swept to the winds by the down-coming from the North of the avalanche of barbarism that they had looked upon with scorn and neglect. They had tried to save the Grecian and the Roman world, without saving the rest of it; and the barbarism overwhelmed and swallowed up the civilization. Take another hint from the life of France, just preceding the French Revolution. Never has there been a period in the history of the world when culture and refinement and art and pleasure were carried to a higher pitch than in the last years of the

French empire. But beneath it was a life looked down upon with scorn and neglect by the rich and the great; and that life by and by rose beneath this fabric that was resting upon it as a foundation, — rose as the flood rises under the ice in spring, broke it into atoms, and whelmed the fair life that was above it, beneath its cold, dark, slimy, and muddy waters. The *sans culottes* were in the palaces and the houses of the great. There was no possibility of saving a few, while leaving the many outcast and neglected.

The atonement, then, of the religion of the present, and that shall be the religion of the future, — the doctrine of atonement that is permanent, that is universal, — must be something capable of taking this living, throbbing, pulsating humanity of ours in its entirety, and bringing it, body and mind and soul, into sympathy with the life of God *right here, now;* not to-morrow, not in another world.

Who, then, are doing this work of atonement? Who are the saviors of the race in this universal sense? Jesus is one of them. For, as I said, the moral teachings and the life and power of Jesus have entered as permanent elements into the religious life of the world; and, thank God for it! they cannot be eliminated. But this doctrine shall not hereafter be confined to Jesus. In their sphere, and according to individual power, individual effort, and individual achievement, it is possible, and it shall be a fact, that every man, woman, and child in the world shall help on the work of universal atonement between humanity and God. The great army of truth-seekers over the world, — those who patiently and persistently, in loneliness and neglect, misunderstood and outcast, day and night, are seeking after truth, — they are helping on such an understanding in regard to God and man, and the relationship existing between them, as shall culminate at last in this perfect work of atonement between God and man. Studying the stars, reading the records of creation as the hand of God has written them in the hieroglyphics on the rocks, that have been laid down slowly, age after age, from the first dawn of the world until to-day; tracing back the record of the life of man, finding out what man is, where he came from, how he came to be what he is to-day; tracing the dim and distant records of the earliest civilization, the experiments that man has tried, succeeded in, failed in, tried again, in the work of bringing to pass a perfect society, — the experiments of man in government, in attempting to bring about the time when the statute-book should represent the really high life and thought and justice of the best men of the world, — this great army of truth-seekers, I say, are first and foremost in this grand world-wide work of atonement. And then there is the army of reformers: not those self-constituted reformers who expect to save the world by passing a series of resolutions, or calling public meetings; but those men who are studying the early facts of the world, who are attempting to adjust the relations between the rich and the poor, between labor and capital; those men who are attempting to bring about a mutual understanding on the part of the alienated elements and classes of society; those men who are engaged in what seemed to be at first purely secular work, — the discoverers, the inventors, of the world, — those men who are studying the laws of nature, who are making them obedient servants and helpers of humanity. Why, some of the purely physical discoveries of the world bear a closer relation to the religious life and the future welfare of humanity, than do some of the great religions of ages in the history of the past. To-day the telegraph and steam are doing more, and they have done more in fifty years, to bring about a sense of universal brotherhood, to help humanity feel that it is one, and that all humanity is related to every other part of humanity, — I say these two agencies, the telegraph and steam-power, have done more to bring about this, than all the preaching and all the churches of the last eighteen hundred years. So the men who are seeking to regulate the sanitary life of cities, to give the poor, the outcast, the down-trodden, an opportunity to live, physically, according to the laws of health, of decency, — these men are doing more than hundreds of those who simply preach and pray, in the work of lifting up humanity toward God, in bringing about such a set of conditions as shall make it possible for those who are lowest in the scale of humanity to climb up into decency and self-respect, and so to find out that they have brains, that they have hearts, that they have spiritual natures that link them to God. And then the great army of witnesses for righteousness and truth, the sufferers and martyrs, those who have had some insight of the divine principles of righteousness, and who have stood for them in the face of mobs and angry crowds; who have stood for liberty, who have stood for the rights of the down-trodden, who have been the martyrs of the outcast races of the world, — those men who have dared to stand in the face of death, while the flames were kindling about them, or, stretched on instruments of torture, have dared to die rather than yield one jot or tittle of truth, — such are the men who are bringing about this grand, universal work of atonement between the human and the divine. And, more than that, there is not, as I have said, one single person who reads this, nor a single human being alive, who may not, in his sphere, become as divine a savior as Jesus, or any of the grandest and sweetest souls of history.

Be faithful and true in your relations where you are placed. Stand by the principles of right as you see them to-day. Be true, be honest, be faithful, be fearless. See what you can of the divine, and embody it as far as you can in your own life, and in the lives of those about you; and you atone, you help bring on that time when the life of all humanity shall be pervaded by the spirit and the life of God.

I wish to call attention to just one thought as my last. I pointed out at the first how a false conception of sacrifice was the central principle in the idea of atonement in the early ages of the world. All are well aware how the sacrifice of Jesus on the cross is claimed as the central idea of the doctrine of atonement, as held in the popular religion; and the idea of sacrifice remains, transformed, taking on a new shape, having a new significance. Sacrifice shall still be the central idea of the world-wide and universal doctrine of atonement, — the sacrifice of the lower to the higher, the sacrifice of ease, the sacrifice of wealth, the sacrifice of whatever be needful to enable you to stand for the highest ideal you can discern of truth and right. Nothing that is worthy of attainment has ever been gained in the history of humanity, and never will be, except by the pathway of the cross, as Jesus used the term, — the cross, in its grand and central significance. If you wish to become rich, it must be by the sacrifice of ease and pleasure, the sacrifice of ten thousand other pursuits that you might like to follow. If you wish for pleasure, you must sacrifice every thing that stands in the way of pleasure. If you wish for fame, for power and influence, you must study, and struggle, and train yourself, and sacrifice, to attain it. And so you find the law universal: whatever you wish to gain, you can gain it only by the way of the cross, — by the giving-up of something that stands in the way of attainment. Look through every nation, every religion, every civilization of the past, and you find it a world-wide truth, that the men who have stood for the highest, who have helped on the progress of humanity, who have lifted men up toward God, have been the men who have been misunderstood, who have been maligned, who have been outcast, who have been lonely, forsaken by friends, forsaken by kindred, out of sympathy with their time, daring to stand alone with God, to sacrifice fame, to sacrifice reputation, to sacrifice money, to sacrifice ease, to sacrifice every thing that stood between them and God. They have been the men who dared to trample pleasure and wealth and fame underneath their feet; marching ahead, the pioneers of humanity, and awaiting praise when, by and by, grateful after-times are willing to write their epitaphs, and honor them on their tombstones. So that still it is true, applying the words to any individual who works for the atonement of humanity, applying them to humanity collective, applying them to Jesus, applying them to any and every one who has in any measure been a savior of his kind, still those words are true, "Surely he hath borne our griefs, and carried our sorrows: yet we did esteem him stricken, smitten of God, and afflicted. But he was wounded for our transgressions, he was bruised for our iniquities; the chastisement of our peace was upon him; and with his stripes we are healed."

XI.

CHRISTIANITY AND EVOLUTION.

IF evolution is true, what becomes of Christianity? Are the two antagonistic, so that one necessarily excludes the other? That depends upon definitions. Prof. Tayler Lewis said, not long ago, that the dogma of the supernatural and instantaneous creation of man, by the fiat of a purely personal God, was the very foundation of Christianity, and essential to its existence. But I utterly fail to see how the two have any necessary relation to each other. So I would answer our opening question by saying, Since evolution is true, therefore Christianity, one of its products, is also true. This statement will hold only of its essential life. Forms and statements may change indefinitely, while the same life that created the forms remains. The life that animates the caterpillar, as it crawls slowly along the ground, is the same life that soars and floats and glitters in the butterfly. The transformation extinguishes not the life. One of the results of evolution cannot possibly contradict evolution itself. I am none the less a Christian, then, because I am an evolutionist. I will even say I am a Christian *because* I am an evolutionist. To justify this position, to trace the progress of religion until it culminates in Christianity, and to show the relation it sustains to those religions that have preceded it, — this is my present purpose.

The first form that the religious thought of the world clothed itself in was fetichism. It sprung up naturally and necessarily out of the best and highest thought of the time. It is nothing to be contemptuously despised, any more than the infant's first efforts at speech. It accorded with all the philosophy and facts then known, quite as well as Christianity to-day harmonizes with the knowledge of to-day. It agreed perfectly with what was known of the universe, and sprang out of that knowledge. It was just as necessary a phase of the religious life of humanity, as twilight is a necessary phase of sunrising. Look at the process of thought out of which it sprung. Man recognized his own personal will and choice as the source of all his movements and power. This was the only kind of power he knew any thing about: therefore, whenever he saw exhibitions of life and power, he could account for them only on the ground of his knowledge and experience. Man can never conceive of any thing that transcends all his knowledge. Thus his experience compelled him to endow with personal will and choice every thing about him. His dreams made him familiar with the thought of spiritual beings existing apart from substantial bodies; and these souls he thought of as the source of life and power. And it was natural for his crude thought, to endow all things with souls. Thus, when a savage buried along with the dead warrior his horse, his bow and arrow and spear, his kettle and his cooking utensils and tent appurtenances, he was not so silly as to suppose that he was going to take these material things with him to the happy hunting-grounds: he believed that the souls of these things would accompany and still serve and be useful to the soul of the warrior. Stones, sticks, trees, reptiles, birds, springs, rivers, — all things were thus alive to him, and represented a personal consciousness like his own.

And, as I have already shown, since the principal forces encompassing early humanity were forces that hurt them, on the action of which they could not calculate, and of which they were therefore afraid, they thought that most of these beings were evil-disposed toward them. They did not know nature enough to control it, and make its powers serve them, and they had not philosophy enough to see how apparent evils might become real good: so, of necessity, their first gods were devils, and the earliest religions were devil-worship. Even to-day, should you go to people in this condition of mind (and fetich-worship is still widely prevalent), and tell them about God, they would perhaps inquire, "But what if this great being should eat me?" and the effect would very likely be, that they would run off affrighted into the jungle, to get away from God.

You will notice here what you will find to be true always and everywhere, that the prevailing thought of

God accords with, and is on the same level as, the prevalent philosophy of the universe, the theory of the world.

The next step in the evolution of religion was to polytheism. This is only fetichism partially generalized. That is, instead of making each tree a god, they rose to the thought of one god for all the trees. Instead of each frog or serpent being a deity, they had thought of a king, or god, of all the frogs or all the serpents. They divided nature off into departments, with a deity supreme in and ruling each. Thus Æolus became god of the winds, Neptune of the oceans and seas, Jupiter of the heavens, Pluto of the under world of spirits, and so on. The individual spirit in each individual thing was not at first destroyed: only there was one chief, like the chief of a tribe, who held the allegiance of all. You see here the social and political condition reflected in the religion. This only shows how all the different departments of human thought and life, social, political, philosophical, and religious, keep even step and progress side by side. One side of humanity does not outrun another. Not only did they have gods of the several departments of nature; but they gradually rose to the conception of abstract ideas, and had deities to represent the departments of thought and the various mental life. So Apollo was the god of eloquence, Minerva the goddess of wisdom, the Muses superintended poetry and the arts.

One step more brought the human mind to anthropomorphic monotheism. That is, they generalized still more; and instead of gods supreme in their different departments, like the kings of sovereign states, they conceived a universal monarchy of the world. But as yet this universal monarch was only a mighty and gigantic being in the image of man; for they could think of no higher power than that which humanity represented. And, though he was supreme, the other gods were not dead. The conception was very much like the theory of our government, in which the president is supreme executive, while still the governors rule, as subject to him, over the several states. Even Israel at first had no higher idea than this of their Jehovah. They sing of him as being "a great king above all gods." The "all gods" are real beings: only Jehovah is "king above" them. In the progress of thought these subject gods at last die out, or else become degraded to the position of rebellious but yet conquered and subject devils. This last was the condition of things in the popular thought of early Christianity; and it is the condition of things in popular Christianity to-day; only that the people who still believe in the devils have forgotten where they came from, and how the belief originated. God is still to most minds only a gigantic man, acting on human motives and impulses, and according to human methods.

But among the higher thinkers of all ages, there has been a tendency to take the fourth and highest step of all, — that leading to a purely spiritual monotheism. Jesus gave it its purest old-time expression, when he said, "God is a spirit," and taught that he did not dwell in any one material place, like the ark, the temple, Gerizim, or Moriah. Modern science, culminating in evolution, rounds out and completes this conception by teaching that God is the spirit and life, underlying and giving form to all things, while yet he transcends all thought, and eludes all attempts to give him shape, and slips through the finest-wrought meshes of all the creeds. This is the perfect monotheism, "God over all, and in all, and through all."

The outward forms of all religions have, of necessity, taken their shape from the prevalent thought of the time. Thus the growth has been from the sorcery and charms of fetichism, through animal and human sacrifices, prostitutions, symbol-worships, idolatries, rituals, and stately ceremonies, up to the high spiritual thought that true worship is the aspiration and adoration of the heart, and that mercy and justice and truth are more than all altars and sacrifices. Only the finest thought and the finest living can appreciate this last even yet. It seems unsubstantial and unreal to the multitude; just as this same multitude cares more for a minstrel than a symphony, for a farce than for Hamlet, for a chromo than a Raphael.

We must now pass to the evolution of the ethical side of religion. Morality and religion are closely connected in our minds; but originally they had no relation to each other. At first men prayed and sacrificed to and worshipped the gods, not because they thought of it as a moral duty, but because they feared them, and wished to ward off some supposed danger; or because they wished some favor, and hoped to gain it in this way. So far was the fetich-worshipper's mind from connecting any moral ideas or thoughts of right or wrong with the worship of his gods, that he did not conceive his gods themselves as moral, or as caring for right. One does not seek to please the devil with righteousness. The only morality of that time grew out of the simple human relationships in which they stood to each other; and the gods and religion had nothing whatever to do with it. The same was and is true of polytheism. In Greece and Rome, the popular religion was in no way connected with the popular morality. The gods were simply the kings and rulers of heaven and earth; and they represented, in their own characters, all phases of human character, even its vices and crimes. They even patronized and protected vices and crimes, so that licentiousness and theft and murder were sometimes a main part of religion. There were, indeed, supposed to be Fates and Furies that punished the more flagrant outbreaks of wickedness; and in later ages they thought of future rewards and punishments. But they were only after-thoughts. Men could even pray to the gods for help in some deed that now would ostracise a man from society, or shut him behind prison-bars.

The moral motive and sanction in communities like ancient Athens were purely social and political; so that, though religion and morality were growing side by side, though they re-acted on and modified each other, they yet sprung from different sources, and flowed like parallel rivers before they come to a junction.

During the early history of Israel, religion and morality

were quite as distinct as among other people. Good worshippers of Jehovah also worshipped Baal and Ashera with obscene rites; they could lie and steal, and be drunken, and murder, could neglect father and mother, and play false to all the moralities of life, and yet not forfeit their allegiance to the national God. But the later conception of Jehovah represented him, through the mouths of the prophets, as a moral God, who demanded "clean hands and a pure heart" on the part of his worshippers.

In the teachings of Jesus, religion and morality came into still closer combination. In his highest thought they became identical. To love God with all the heart, and the neighbor as ourself, became at once the summing-up of all religion, and, at the same time, the perfect formula of morals. But in the hands of his immediate followers, and of those who gave shape to Christianity as an institution in the world, religion and morality became separated once more. Religion was metamorphosed into a "scheme of salvation." The thing to be saved from was not unrighteousness, but hell; and the means of salvation were not right thinking and living, but baptism and faith. So the grand work of Jesus was partly undone by his disciples. Indeed, his thought was so far beyond his age, that only a few could comprehend it, and it was necessarily degraded to the level of the common intelligence. The Church became a close corporation, holding the keys of heaven as a corporate privilege, for the exclusive use of its members. And one "without," however moral and godlike, could only find refuge among the devils in hell; while those "within," though ever so vile, had become partakers of the corporate spirit and grace, and so had gained a "right to the tree of life, and could enter into the city." This is Romish doctrine still. The brigand of modern Italy looks to his saint to help him plunder, and returns the favor by offering at his shrine a saint's share of the booty. So it is nothing strange that Tetzel should travel through Europe selling papal indulgences, the privilege beforehand of committing all sorts of sins and crimes, so only that he recompensed the Church, which, "for a consideration," promised to keep the gate of heaven ajar for him till he got ready to go in.

And this same idea taints Protestantism still. The ecclesiastical conditions of salvation are held in such high esteem that even to-day a heresy is less easily forgiven than an immorality. Mr. Moody stands up in New York, with all New York Protestantism at his back, and, without one audible undertone of dissent, asserts that morality "don't touch the question of salvation." It is still only an ingenious ecclesiastical "scheme," by which God is enabled to snatch sinners with his one hand, called salvation, out of his other burning, dark hand, called damnation. If only church-members are logical, it is no special wonder that Washington should be full of "Christian statesmen" who dabble in "Credit Mobiliers," and that ministers like Winslow should find it convenient to go to Rotterdam. It is not enough that the Church should preach that its members ought to be moral, so long as it also preaches that its great end and aim, salvation, hangs on something not connected with morality.

Evolution goes back, and, taking up the pure word of Jesus, completes and enforces it with all the knowledge and emphasis of modern science. When it makes all law, physical, intellectual, and spiritual, only the manifestation of the living, loving, righteous God, then it absolutely identifies religion and morality, or makes morality only a part of the greater and grander thing, religion. God is not any longer a being sitting apart, to be pleased or displeased by what you do, or do not do, as pertaining personally to him, while life here in the world is something disconnected from him. But God is here, all about us and in us. He is in sun and air and ocean and earth. He is in heart-beat and brain-throb, in every fibre and muscle and thrilling nerve of the body. He is not only in the truths of religion, but he is in the truths of science. The laws of the intellect are his laws; the light of truth is his light; the moral relations in which we stand to our fellow-men are the expression of his thought and life in humanity. So that duty to God becomes absolutely identical with all human duty. Righteousness before God is absolutely identical with all human righteousness. Pleasing God is obeying all his laws. Salvation can possibly be nothing more nor less than coming into perfect accord with the whole life and movement of things. A man is saved just in so far as he knows and obeys the laws of God. Perfect salvation is perfect knowledge and obedience. A man in this condition is of necessity perfectly moral and perfectly religious.

Thus the teaching of evolution completely accomplishes what Jesus began, but what popular Christianity failed to carry out, — it identifies and makes one the moral motive and the religious motive. The gods have become one; that one has become spirit; and that spirit has become the life and inspiration of all goodness and truth and beauty in state, in society, in the individual, in art, in letters, in science. Indeed, in the grandest sense, God has become "all and in all."

If evolution be true, the life of the universe is one life. And since religion is a part of this life, if evolution be true, the religious life of the world must be one. Let us, then, search for the essence of religion, see whether it exists everywhere, and, therefore, whether there are many different religions, of which Christianity is one; or whether there is only one religion, of which Christianity is the highest outgrowth and expression. Science teaches us that there is only one life on the globe. From the little viscous globule that palpitates in primeval seas, or the lichen that creeps over the rock, up through all ascending forms of plant and animal, till you reach the infinitely involved brain of Newton working a problem in the differential calculus, the imagination of Praxiteles seeking the hidden god in the block of marble, or the complex arts, societies, and politics that issue in our world-wide civilization, — everywhere and all through and all up, it is one life that beats in it all. So evolution teaches that the blindest gropings in the realm of religion were only poor, weak human hands and feet feeling after the lowest step in "the world's great altar-stairs, that slope through darkness up to God." The fetich-worshipper, with the best light and knowledge he had, went "feeling after God, if haply he might find him," who is "not far from any one of us." It was not depravity on his part when he stood in awe of, and worshipped, and sacrificed and prayed to, a stick or a reptile. He saw therein the infinite mystery and life of the world, and interpreted it as well as his ignorance permitted; and so constructed a religion as high for him as ours is for us, — that is, a religion as true and high as his philosophy enabled him to think. He was not so foolish as to worship the stick: it was the mysterious and wonderful life in the stick, he adored. He who to-day reverences the church, the altar, a day, or the Bible, is doing the same, with the difference there is between a man and a child. The child thinks and feels and imagines as a child; but it is the same life, the same brain, the same heart, that make the after man. So that which the child-man reached after and thought of was the best interpretation he could give of the same God in nature that "the heavens declared, and the firmament showed, and day and night uttered speech" of, to the Psalmist, — the same that we see in all things to-day. And when, in after times, they offered to God their children in human sacrifice, it was not human depravity that prompted the (to us) murderous deed. Jephthah tenderly loved his daughter; Abraham's heart yearned over Isaac;

the Ganges mother clung to her babe with the same heart-wringing mother-love that throbs in your bosoms. But the gods were inexorable; and they gave with heart-break and tears that which was dearest, because heaven demanded the precious gift by all the sanctions of religion. Though the deed was horrible, it was reverence and fear for the gods, that nerved the arm, and steeled the heart. It was the same religious motive in the heart as that which made the martyrs, and has been the inspiration, in all ages, of heroism and noble deeds. And even the services of impurity had behind them the motive of supposed divine sanction and religious obedience. So that when the Santal uses his fetich for a charm, when the Parsee fire-worshipper bows before the rising sun, when the Hindoo lies down to be crushed by his Juggernaut, when the Chinaman offers incense before the image of his ancestors, when the Buddhist devotee sits beneath a tree, and all day recites some holy and magic word, when the Romanist bows to the crucifix, when the Ritualist recites from his book, when the Quaker sits silent and waits for the spirit, when the Protestant bows in prayer, or when the philanthropist goes out on some deed of mercy, believing that "he prayeth best who loveth best," — in all these human bosoms beats the one throbbing, human heart; and each, according to his knowledge, seeks to worship the mysterious life of all things, that we call God, after the highest and most sacred fashion which he has learned. However inadequate his conception of God, he still seeks to find the

> "Father of all, in every age,
> In every clime, adored, —
> By saint, by savage, and by sage, —
> Jehovah, Jove, or Lord."

The thought of God that is held may be wild, chaotic, and, to our mind, wicked. But his art and his government and his civilization are the same; and he thinks, in the one case, on a level with his thought in the other. Religion can have no expression higher than the age is capable of: so it advances with the advance of man. But the point is, that all religions are reaching out after the same God that each one of them fancies it has found. It is the same religious nature of man in all the varied manifestations. Thus religion is one just as art is one, or literature is one. Each age and each nation finds its own expression; and each one is true or false according as it approximates more or less nearly to the perfect truth of things.

But when I say that religion is one, do I then say that one religion is as good as another, and that Christianity is no more divine than any other? Let us see. Human thought is one; that is, it is all alike the product of human brains, and differs only in quality and degree: do I therefore say that Newton's thought is on a level with, and no better than, that of a clown? Poetry is one: is therefore Shakspere no higher or more divine than Tupper or Walt Whitman? Art is one: is this to say that the snow-man that makes the holiday frolic of a lot of boys is to be put on the same level as the Moses of Angelo, or the Christ of Thorwaldsen? In all these cases, the infinite range of production is the work of precisely the same faculties, that differ only in the kind and quality of work. And each special work is true and valuable just in so far as it expresses the true and the divine in its department. So all religions are the outgrowth of the divine, the religious nature, in man. But there are all degrees of truth, of excellence, and therefore of divinity, ranging from the deformities of the rudest fetich, up to the divinest thoughts and scriptures and hymns and prayers of the loftiest seers of the world. Christianity, then, is not something thrust into, but apart from, the growing order of the world, any more than Handel's "Messiah" is out of tune and accord with all the music of humanity that, beginning with the rudest song and pipe of reeds, leads up to and culminates in its grand harmonies. Christianity is the highest outcome of religious evolution, just as man is the highest outcome of animal life. It is no more severed from the rest than the full-blown rose is severed from the little twig that broke the seed, and developed into the bush. It is the bright-colored, sweet-scented, and consummate flower on the topmost bough of the religious life of man.

Since, then, Christianity is the result of evolution, is it to be expected that evolution will still go on, and ultimately outgrow and leave Christianity behind? This is the hasty logic of some. Christian schemes of salvation, Christian cosmogonies, Christian ecclesiasticisms, Christian rituals, — these may and probably will be outgrown.

Much of what is called Christian theology will be sloughed off and left behind, as a growing bean rejects its pod. All life takes to itself form, and clothes itself in outward symbols and institutions; but these are only clothes that are cut and worn after the fashion of the age. A boy may change the cut and style of his clothing with every year; and he must put on larger as he grows, to suit the developing size and figure of his body; and he may wear as many fashions as ingenious tailors can invent; but all through, from infancy to age, it is the same boy becoming youth and man. If, therefore, Christianity does put off its old clothes, and put on larger ones as it gets larger, it will not necessarily follow that Christianity itself will be outgrown and left behind. If it contain in itself any touch of the universal and eternal, it must live forever; and if this something that is eternal in it be of its very essence, as Jesus taught it, we may still logically hold that it is Christianity.

As I have already shown, when Jesus said, "Thou shalt love the Lord thy God with all thy heart, and thy neighbor as thyself," he gave utterance to words that absolutely identified religion and morality, and linked all life with God in the same sense in which it is to-day done by the doctrines of evolution. These words Jesus made central; and they are the formula of a perfect human life. So that, so far as we can see, the utmost progress of evolution in human life — individual, social, political — can only approximate more and more closely to this infinitely progressive and expansive ideal. It means the command to be perfect; and even God himself cannot outgrow perfection.

While, then, the popular forms and creeds and institutions may be outgrown, and replaced by better, the essential spirit and life of Christianity will become more and more the essence and spirit of evolution itself, so far as it bears on humanity. The whole force of evolution, henceforth, will lift up and urge on humanity toward the fullest and highest Christian life.

Is evolution, then, a radical or a conservative element in religion? It is both. It is radical in so far as it eats away, tears down, and leaves behind, the transient and perishable forms; for these things, when decayed and fallen, only become obstructions in the pathway of human progress. But so far as essence and life are concerned, evolution is conservative. It teaches that the one, all-important thing is life. The forms live for it, and not it for the forms: so it does not regard a partial, incomplete, or even grotesque form, so objectionable that it is to be got out of the way at so great a cost as the risk of the life it holds. It will keep a flower in a cracked flower-pot, rather than risk the flower itself, by knocking the pot away. It will leave a man the inadequate shelter of his hut, rather than tear it down about his ears in a storm, before it can invite him into a better house. Regarding all things as a growth from small beginnings, it does not teach the top of a tree to ignore its roots, nor a mansard house-roof to despise the mudsills, nor a man's head to scorn his feet. It is, therefore, tolerant of ignorance and half-development, and the slow process by which a little comes to be more. Since it took God millions of ages to get the earth habitable even for reptiles, it does not ask him to lift the reptiles up into humanity, and make humanity absolutely complete, in six days.

Thus evolution is tolerant even of fetichism. It would not knock an idol out of an ignorant religionist's hands, except as it can replace it by a purer and truer symbol. It would not destroy Mohammedanism, except by replacing it in the minds of its devotees by a nobler God than Allah, and a better morality than that of the Koran. It will permit the Romanist Irish girl to keep her beads and her mass until she can grow into a conception of a higher and more spiritual religion. To take these away, and leave her empty-handed and empty-hearted, is against its whole spirit, and seems to it impiety. It lets the boy keep his toys until, having become a man, he is ready to "put away childish things." So the attitude of evolution toward orthodoxy is such as makes it rejoice that men will cling to orthodoxy until they can see and feel that something else is better. The moral and religious life *first* and *highest*. Religious forms and creeds and rituals are only the expression of its life, and made to serve it. As the life lifts and broadens it will lift, break up, and sweep away its covering that threatened to be its bond, as the spring freshet tosses on its bosom and sweeps away the ice that covered it, but was not strong enough to hold it in chains.

Evolution, then, in religion, will seek to find and spread

abroad all truth, believing that the inner life is able to stretch the bark, and fit it to its various stages of life and growth. It tolerates the twilight because it leads to the sunrise. Its trust is in time, light, growth, and God. It is patient with the lower form of life, and also with the lower life of man; for it knows that these are but the necessary childhood of life, that at last shall grow up to and culminate in

> "The crowning race
> Of those that, eye to eye, shall look
> On knowledge; under whose command
> Is earth, and earth's, and in their hand
> Is nature like an open book;
>
> "No longer half akin to brute;
> For all we thought and loved and did,
> And hoped and suffered, is but seed
> Of what in them is flower and fruit."

And so, holding always any present to be but the seed of a better future, it trusts in and waits for the still larger unfolding of

> "That God which ever lives and loves,
> One God, one law, one element,
> And one divine, far-off event
> To which the whole creation moves."

XII.

IMMORTALITY.

The belief in some kind of a future life seems to be as old as humanity. The testimony on this subject is much more ancient than any written records. It speaks to us from the excavations where have been discovered the remains of prehistoric man. So the old question of Job, "If a man die, shall he live again?" had been asked by the same human heart thousands of years before his day; and, rightly or wrongly, it had been answered in the affirmative.

So far as we know, the animals are not perplexed over the problems of life; they do not try to settle right and wrong. They shelter themselves from storm; they bask in the sunshine; they eat and drink and sleep; and, when the hour of death approaches, the sheep or horse or dog lies down with no anxious questioning as to what may come after death. It never pauses on the edge of destiny, like Hamlet, to soliloquize: —

> "To die — to sleep, —
> To sleep! perchance to dream: ay, there's the rub;
> For in that sleep of death what dreams may come,
> When we have shuffled off this mortal coil,
> Must give us pause."

Such thoughts and speculations as these are confined to humanity.

At what precise point in the upward lifting of animal to man, the newly developed human heart began to raise these questions about a possible future, it is now impossible to tell. But we can go back far enough to trace the probable course of reasoning out of which the belief first rose. Indeed, we may find this same primeval belief and primeval reasoning in existence to-day. If you wish to see how a century-old oak-tree sprouts and grows, all you have to do is to look at a bursting acorn, and watch the unfolding of the tiny stem. So the thoughts of the world s childhood can still be read in the mental processes of those races that are in their childhood still. In this way we can come at the origin of the world's belief in a future life.

These first men reasoned well, considering their knowledge and their mental powers. It seems to have been something in this way. They looked on the body of some one who had died: here were the feet that were so swift in the chase, or on the war-path; here, the hands that bent the stiff bow, or hurled the spear; the eye, that was quick and sure as the hawk's for his prey, was dull and sightless; the heart beat no longer; and the ear that was never deaf to the shout of a comrade or the taunt of an enemy was now equally indifferent to both. And they said within themselves, as they saw him stiff and cold and still, "This is not all there was. He who loved us, and fought our enemies, and hunted with us the common prey,— he is not this body: he has gone away, and left it. He was more than this, and different from it." Thus, of necessity, there grew up the thought of life, as separate from the body.

Other lines of reasoning led to the same result. They saw forms and heard voices in their dreams. Knowing nothing of their true origin and explanation, of course they must explain them as best they could. They saw no reason why their dreaming experiences were not as real as their waking ones. So these shadowy forms that came and hovered about them, and talked with them, they learned to look upon as real persons. And when, after a friend had died, he came to them in dreams, this seemed to solve for them the mystery of death; for here was the same old face and figure smiling upon and speaking to them, though the body was in the grave. So they were compelled to believe that every person was double, having a body and another self, a spirit or shadowy one. They knew nothing of the laws of sunlight, by which every object casts a shadow when the light falls on it; and they saw that at times, when it was cloudy, or when they went under the trees or into their huts, these mysterious shadows went away, or hid themselves. Thus they began to identify their shadows with this second self that came and went in dreams, and that went away and left the body entirely at death. It was this second self that looked up into their faces from the surface of glassy lakes or springs. So real to their minds was this belief, that they held it dangerous to walk along a river's bank where their shadow would fall into the water, lest a crocodile should seize it, and so cause their death. They learned to believe that all dreams were real spirit life, their souls going off on a journey, or else other souls coming to visit them. From this grew the world-wide belief in visions as revelations from the unseen world. And all the modern superstitions about dreams as being warnings, or signs, have had a similar origin, and are only survivals from these old times.

They identified the soul with different parts of the body. It was the pupil of the eye, or the heart, or the blood, or the breath. In Hebrew thought, God was supposed to have breathed the soul into the nostrils of Adam; or, as in the law it is said, "the blood is the life." In Homer, the "shade" or soul of the warrior rushes out through the wound that causes his death. Among other peoples, they talked of the heart's going away, or coming again, as the life ebbed or flowed.

So much, then, by way of explanation of the way in which the thought of a dual life in man first sprung up. A word now as to the nature and location of this future life from the first, until to-day. The earliest notion seemed to be that the soul of the dead would still continue to love its old-time body and place of abode. So it was supposed to hover about the spot of burial, or wander through the village that was its home. This oldest of all ideas still lingers in the thrills of superstitious fear that many yet feel in passing a graveyard at night. It is the survival of the ancient thought that the spirit still stays about the grave. And the stories of haunted houses are only remnants of the old-world belief that the soul still clings to its earthly abode. And the Egyptian custom of preserving the body in the form of a mummy is not only connected with their belief about a bodily resurrection, but also grew out of the thought that, so long as they kept the body, they could also prevent the soul from going far away.

The next stage of belief is higher. It was thought that the ancestor of the tribe was still interested in it, and retained his authority over it, after he had passed away. So it was supposed that he was engaged in preparing a home and kingdom for the spirits of his descendants in the other world. They believed that in this kingdom he received and ruled over the souls of his tribe as they entered the shadowy land. Dying, then, to them, was simply going to the fathers of the tribe.

A third stage in this belief was that which is represented by the Elysian Fields of the Greeks and Romans, the Islands of the Blest, or the Halls of Odin, as they were imagined by our Scandinavian forefathers. Socrates, on the eve of his death, talked with his disciples of his expectation that in the Elysian Fields he should meet and converse and walk with the wise and the great and the good of the olden times. And he so separated his body from himself, that, when one of his friends spoke of his burial, he said playfully, "You may bury me as you please, if you can catch me." *He* expected to be away, though the body was left behind.

The Jews at first had no knowledge of any future life; but, when the notion of the Messianic kingdom grew up among them, they believed that the coming deliverer of David's line would raise the dead, and establish an ideal and everlasting kingdom on the renovated earth. So their hope was purely a material one.

I would have you take notice that thus far there is almost no trace of any friendship with the gods, or of any heaven to be spent in their company. The Halls of Odin are really no exception to this statement; for this eternal feast was only for the bravest of the warriors who died fighting in battle. Common humanity had no part in it.

The last historic step of which I wish you to take notice is the early Christian conception of heaven with God. This, at first, was only the Hebrew Messianic idea somewhat modified. John, in the Apocalypse, pictures "the new heavens and the new earth," and describes "the city of God descending from God out of heaven," to abide on the earth. In this gem-gated and gold-streeted city, the "nations of them that are saved" were to "dwell; and God was to dwell with them, and be their God." Reinhabiting their resurrected bodies, the redeemed were to live in this material and glorious city. They would need no sun nor moon, for God was to be their light. They were never to go out, and death was never to enter. Such was the immediate expectation of the early Church. But, as this delayed, they began to imagine the city of God as just above the solid arch of the firmament, and to think that there their friends were gone. When modern astronomy destroyed this old-time firmament, and the world learned that it was only an optical illusion, of course the location of heaven must be changed. (For now we know, though it is very recent knowledge, that the arch of heaven is only light and air, like that which is all about us.) Then speculation placed heaven in the moon. By others, books have been writted to prove that it is in the sun. "The To-Morrow of Death" twists science in the most fantastic way, to support this last idea. Some have thought heaven might be on the star Alcyone, which is supposed to be the centre of the great star-system to which our solar-system belongs. And in a recent sermon Mr. Talmage (who talks as if he had been there, and knows) asserts that it is located on the central sun of the universe, which he supposes to be millions of times larger than any other. Since Mr. Talmage has given the weight of his tremendous authority on this point, of course it is of no practical importance for me to add that no other man living knows any thing about the existence or the nature of any such orb. We know that there are stars so distant that the more than lightning-like velocity of light requires some millions of years to travel the distance that separates them from the earth. But though Mr. Talmage's heaven is farther off than these, he assures us that souls can make the journey in a fractional part of a second. This is a great comfort ; for otherwise we might be left to fear that even Adam, during these six thousand years, had hardly got started out on his journey even yet.

One of the standing charges of the Church against science is, that it is materialistic. I wish, in passing, just to call your attention to the fact that the whole ecclesiastical conception of the future life has been, and is still, pure materialism. The material body is to rise and dwell in a material heaven.

Such, then, have been some of the more prominent speculations of the world concerning the nature and location of heaven. We can think and believe and hope; but any intelligent and thoughtful man will hesitate before he will assert that he *knows* any thing about it. Ignorance is confident; for it has always been true, as Pope says, that "Fools rush in where angels fear to tread." But reverent knowledge will wait for light.

All these forms, then, under which the hope of a future life has been held, have been disproved and outgrown by modern knowledge. And here comes in the great question, Must we therefore give up the hope as irrational? Is it only a dream of the world's childhood from which its larger-grown intelligence awakes? As a hope, it is most magnificent; as a dream, it is beautiful and grand. And it argues strange and high capacity in man, that he should even hope or dream of such a thing. Of course we will give it up if we have to; but most certainly we cannot be expected to till then. Let us, then, raise the question as to whether there is any theory left that an intelligent and rational man can hold. As I look over human thought, I find two, which I must try and make clear to your minds.

(1) There is, then, first, a materialistic theory which

our present knowledge cannot disprove nor make absurd. It is well known that all our senses have only a certain narrow range within which they are able to bring us into sensible contact with the world about us. All outside this range we are unable to reach. For example, we do not see all forms and colors; we do not hear all sounds; we do not smell all odors; we cannot consciously touch all substances; we cannot taste all flavors. Vision depends on the wave-motions of light. If these motions are less than a certain number in a second, they do not produce on the eye the sense of vision: if they are more than a certain other number, we cannot see them. Thus the narrow range between two definite numbers that represent the quantity of wave-motion in a second, is the limit to our sense of sight. A whole world of things may lie on the one side, and another world on the other side, of this limit, in the presence of both of which we are totally blind. So there are forms and colors all about us on every hand, that we do not, and can not see. And a similar thing is true of our ears. We can hear only within certain definite limits. Were our senses acute enough, the silence of a summer midnight would become to us a thunderous tumult. We could hear the flowers grow in our garden until the stillness broke into a noise as loud as the waves on the seashore in a storm. Huxley tells us that, if we could hear the movements in the growth of a stinging-nettle, it might become to us as loud as the rattle and roar of a great city.

For any thing we know to the contrary, then, a refined

and spiritualized order of existences may be the inhabitants of another and an unseen world all about us. Milton has said,—

> "Millions of spiritual creatures walk the earth
> Unseen, both when we wake, and when we sleep."

Of course, the poet's words do not prove this true; and all I care to say about it is, that such a thing is possible. That we do not see or hear them, is no proof that they do not exist. The inter-planetary spaces may be the home of a universe of life, for all our senses can say to the contrary. I suppose that some such idea as this lies at the base of modern Spiritualism. I have never yet been convinced that it is proved; but certainly any knowledge we now possess cannot say that such existence is impossible.

A remarkable book has been recently published, called "The Unseen Universe." It is the work of two prominent English men of science. It attempts, on a scientific basis, to establish the possibility of the existence of an unseen world, closely connected with this, from which the visible universe came, and to which it will return again. It would take too long for me to give you the outlines of their argument. They do not prove the fact; but they do prove the possibility. And this is all I care for now.

It is not unreasonable, then, to believe in the possibility of another life, even if your theory of it be only a refined form of materialism.

(2) But there is another possible theory that is purely

spiritual. Ever since the first man saw his shadow, and talked about his other self, it has been common to speak of man as a combination of mind and matter. To find out what mind is, whether it is a product of matter, or something distinct from it, has been the study of all ages. But the question is not settled yet. Some scientific materialists claim that they have settled it, that thought is a product of the brain, just as bile is a product of the liver. Moleschott — as though the saying decided the question — declared, "No thought without phosphorus;" but that hardly proved that thought was the product of phosphorus. No flash of lightning without the proper atmospheric conditions; but that does not prove that these conditions create electricity. What, then, is the belief of the best modern thinkers, about the relation of mind to brain? It is inclined to believe that the only real things that exist are the mind and God, and that the universe is only the infinitely varied manifestation of God to the human consciousness. For instance, let us see what it is that we know about a desk. I touch it with my hand, and feel the sensation of touch. Now, all I really know is my own sensation, and that something outside of me has produced this sensation. But that all this outside something that I call the universe is any thing other or more than the manifestation to me of the infinite God, I do not know.

And so far from mind's being explained as the product of the brain, all we know is, that the action of mind coincides with certain molecular movements in the brain.

But all these movements can be explained and formulated without any reference to the mind at all. The movement of electricity along a telegraph-line is accompanied by certain molecular changes in the wire itself; but the wire is not electricity, neither does it produce it. Thus modern science has found it utterly impossible to explain mind either as a part or a product of matter.

It is perfectly reasonable, then, for any man to believe in a purely intellectual and spiritual existence, apart from any material form or substance.

You will notice that I have not claimed to prove either the materialistic or the purely spiritual theory of a future life. My only purpose has been to show that there are theories that intelligent people can hold, even in the clearest light of modern science, without laying themselves liable to the charge of being irrational.

Having this basis, then, to stand on, let us review some of the probabilities of the case. No intelligent man, I suppose, will claim that he can demonstrate immortality. The most we can do is to weigh the probabilities, and see how strong a foundation we have on which to build our hopes. What are some of the stones, then, that go to the making of this foundation?

(1) I wish to make one negative point. Our present ignorance concerning the nature or even the fact of an immortal life is no valid argument against its reality. Humanity knows nothing beyond the range of the experience of humanity. By the very terms of our supposition, the immortal life is something above and beyond the

earthly experience of man. The caterpillar probably knows nothing about any life higher than that of his toilsome crawling on the ground; but that is no proof against the fact that we know he is to become a butterfly. The boy knows nothing about manhood, and cannot know. Though he sees men and their labors all about him, he has and can have no conception whatever of what it means to be a man: it transcends all his experience.

So, if there is a life very much different from, and very much higher than, our present one, it is not strange that we are ignorant of it. It is perhaps impossible that it should be otherwise. I could not, with all my trying, make you understand any thing entirely unlike all you have ever seen or heard. So, if an angel should come and tell of another life, it would mean nothing to us, unless he could translate it into terms of our own experience. We could not understand a "light that never was on land or sea." Our ignorance, then, is not even a probability against the belief.

(2) I ask you to notice that the belief exists, and has the field. It holds the position, and will stay, unless dislodged; and, of right, it ought to stay. It is practically true to say that all men everywhere have believed in a future life: no matter under what form, the fact remains. The exceptions have been hardly enough "to prove the rule." The burden of proof, then, lies with the doubters. If the universal belief is a falsehood and a cheat, it is for them to prove it so. The universal human instinct and longing is well uttered by Tennyson, —

> "Thine are these orbs of light and shade;
> Thou madest life in man and brute;
> Thou madest death; and lo! thy foot
> Is on the skull which thou hast made.

> "Thou wilt not leave us in the dust:
> Thou madest man, he knows not why;
> He thinks he was not made to die;
> And thou hast made him: thou art just."

No man can call me unreasonable for holding the world-wide creed until he has proved to me that it is not true.

(3) As the process of evolution goes on, life grows fuller and more intense. And the more intense the life, the stronger grows the belief in its continuance. There are times of sorrow and weariness when we feel that life itself is a burden, and that the only real rest and peace are to be found in long and dreamless sleep. I have heard of some who felt tired at the thought of living forever, and who hardly wished for the certainty of such belief; and it is fabled that old Tithonus begged the gods to take from him the gift of an earthly immortality, it grew to be such a burden. But, in all these cases, the weary ones cumber their thought of immortal life with the burdens of endless earthly cares. So it is not life that men grow weary of : it is only the troubles and sorrows that take away from the sum and fulness and power of life. Life, of itself, is always joy and strength. So, in spite of suicides, of the poetry that longs for the grave,

of the weariness that would even push away the proffered cup of everlasting life, it is still true that

> "Whatever crazy Sorrow saith,
> No life that breathes with human breath
> Has ever truly longed for death.
>
> "'Tis life whereof our nerves are scant, —
> O, life, not death, for which we pant;
> More life, and fuller, that we want."

If it were true that, the larger and grander life becomes, the more nearly it seemed to culminate and reach its completion, there might then be room for the thought that some day it would reach its limit, and so naturally and contentedly come to an end; but just the reverse of this is true. It is the narrow and stinted life that thinks it knows all, and is contented. It is men like Newton that talk of standing and gathering a few pebbles on the shore of an infinite ocean that lies all unexplored before them. The larger life and knowledge grow, then, the more they reach out and hunger for the infinite still unattained.

(4) It is a law of evolution, that, when it reaches the highest form, its whole force, which destroyed lower forms in the interest of higher, now turns to the perfecting and continuance of this highest form. It climbed through lower animals to man. But in man the highest physical form is apparently reached. So now it works only to perfect and continue humanity. It creates no

new brain, but only lifts brain higher. It creates no new moral life: it only lifts and continues and intensifies the old moral life. It makes no new mind or spirit: it only broadens and deepens mental power, and increases the consciousness of the divine and eternal. The highest result of evolution, then, is to increase and strengthen in man his consciousness of the divine and spiritual in life, — the beauty, the truth, the right, the ideal, the eternal. He comes thus into conscious possession of things that must be a part of the everlasting life of God.

To many men, poetry is nothing. It has no real or appreciated existence for them. They have developed in themselves no poetic faculty to which it can appeal. But to Milton or Dante it was the most real and intense of all facts. And they, and such as they, prove that it is a part of humanity. There are many for whom music is no reality; but Beethoven and Mozart ask for no demonstration of its life and power. Their very life was music. Praxiteles and Angelo, and Titian and Powers, need no proof of the reality of art. To such men, the universe is art. And you will notice that men become conscious of poetry or music or art, only as they develop and live in those faculties of their being that find expression in these. Why, then, is it not reasonable to suppose that the great religious masters of the world so developed the spiritual, the divine, the eternal, in themselves, as to become conscious of these things, as lower men are conscious of the material? Emerson says, "I admit that you shall find a good deal of scepticism on this subject, in the streets and hotels, and places of coarse amusement. But that is only to say that the practical faculties are developed faster than the spiritual. Where there is depravity, there is a slaughter-house style of thinking." That is, the man who lives in the spiritual and eternal believes in the spiritual and eternal.

(5) The man who denies immortality must explain why it is, if it is not true, that men possess the thought, the hope, the dream. The most flitting fancy has its cause as much as has a mountain. And it is the fundamental doctrine of modern science, that human faculties are the result of outside forces that created them, and to which they respond. For example, the existence of eyes proves light and objects to be seen; for it is the play of light and external objects of vision, on the human organism, that made the eyes. Sound created ears; and so the very existence of ears proves the reality of sound. The sense of justice, of beauty, of truth, can thus be explained only by supposing some external realities that made them, and which they thus represent. So the most natural explanation of these hopes and longings after immortality is, that they are created by, and that they represent, some eternal reality from which they have sprung.

(6) No creature can think beyond himself. The art of the world proves that art is a native element of humanity. The poetry of the world proves that humanity is capable of poetry. A thousand martyrs prove that heroism is a part of humanity. That man, then, can

think of God and the infinite, proves that there is something of the divine and the infinite in man. If a horse could sit down and meditate; if he could study his own structure, scan the universe, put noble thought into noble verse, think and speculate about the nature and destiny of horses, — it would be held to prove that he had capacities that lifted him out of the plane of the equine, and gave him brotherhood with the human. If man, then, can think and study and speculate beyond his present self, it indicates that there is in him the possibility of overstepping his present limitations, and emerging upon a higher plane of existence.

(7) Then there is the sense of justice, the imperishable belief that somewhere and somehow all things shall come out right. We are perpetually pained here with the sense of wrong. Like the Psalmist we cry out that "all the foundations of the earth are out of course." The wicked prosper; and good is persecuted. The problem of the Book of Job is the problem of all the world and of all time. But Job did not answer it; and this world never has answered it. What is the end of all the sorrow and all the wrong? If there be no justice and right at the heart of things, then whence came this human sense of just and right? It must be the response in us to some eternal reality. And, if just and right do represent an eternal reality, then all must some time be well; we must

> "Trust that somehow good
> Will be the final goal of ill."

But we see it not here and now; and so, out of this longing, men have always built a better future. And to the question, —

"What hope of answer or redress?"

has always come the response, —

"Behind the veil, behind the veil."

What, then, is our thought? The belief in a future life is a natural and an universal one. It may claim the credit of being native and essential, unless it can be disproved. It cannot be disproved. The most that doubt can do is to say it does not know. It may stand, then; and no one may justly charge it with unreason. Beyond this there are many indications that point toward this belief as their most rational solution. This hypothesis of a future is the one that most naturally accounts for all known facts. Such being the case, we may as logically claim it as the astronomer claims a new planet, as yet unseen, as the needed explanation of the perturbations and movements that ask for some such cause.

The most important thing for us to consider practically is the work of personally co-operating with those forces and tendencies in us that are fitted to lift us up into vital relationship with the spiritual and the eternal. Progress sloughs off, and leaves behind those things that are not fit to endure. We can make no better preparation, then, for the future, than to develop in ourselves so full and noble a life that God cannot afford to lose us.

Let us make ourselves a part of the permanent good of things, a portion of the eternal order. Then, because that lives, we may live also.

As illustrating how out of darkness comes grander revelations than day could make, and as indicating how the truth may be a better one than many doubts and fears would sometimes indicate, I cannot close better than by quoting Blanco White's sonnet on "Night and Death:"—

> "Mysterious night! when our first parent knew
>   Thee from report divine, and heard thy name,
>   Did he not tremble for this lovely frame,
> This glorious canopy of light and blue?
> Yet 'neath the curtain of translucent dew,
>   Bathed in the rays of the great setting flame,
>   Hesperus with the host of heaven came,
> And lo! creation widened in man's view.
> Who could have thought such darkness lay concealed
>   Within thy beams, O sun! or who could find,
> While leaf and fly and insect lay revealed,
>   That to such countless orbs thou madest us blind!
> Why do we, then, shun death with anxious strife?
> If light can thus deceive, wherefore not *life?*"

www.ingramcontent.com/pod-product-compliance
Lightning Source LLC
Chambersburg PA
CBHW011600170426
43196CB00037B/2912